Fragile Ecosystems

Fragile Ecosystems

Evaluation of Research and Applications in the Neotropics

Edited by
Edward G. Farnworth and Frank B. Golley

A Report of The Institute of Ecology (TIE)
June, 1973

Springer-Verlag New York • Heidelberg • Berlin

Edward G. Farnworth and Frank B. Golley
Institute of Ecology
Rock House
University of Georgia
Athens, Georgia

Library of Congress Cataloging in Publication Data

The Institute of Ecology.
Fragile ecosystems.

1. Ecology—Tropics. 2. Ecology—Latin America.
I. Farnworth, Edward G., ed. II. Golley, Frank B.,
ed. III. Title.
QH541.155 1974 574.5'098 74-8290

Printed in the United States of America

ISBN 0-387-06695-0 Springer-Verlag New York • Heidelberg • Berlin
ISBN 3-540-06695-0 Springer-Verlag Berlin • Heidelberg • New York

FOREWORD

The Institute of Ecology (TIE) was organized to provide a mechanism for addressing ecological and environmental issues that were beyond the special interests of ecology as a profession. One method of evaluating such issues is the workshop, and this report describes the results of the third TIE workshop on a major environmental subject. The ecology of tropical regions is of interest to all the inhabitants of the biosphere. The tropics provide mankind with both the opportunity for and the challenge of essential resources, land for settlement and development, and waters for numerous uses. Moreover, they provide examples of misuse of the landscape, fragility of ecological systems, and serious environmental problems. Unfortunately, the study of the ecology of the tropics has not kept pace with the ecology of other regions. The purpose of this report, therefore, is to determine the research approaches that will lead to advances in our theoretical understanding of tropical systems and, more importantly, in the application of that knowledge to human problems. Although the principal focus of the report is on the neotropics, it will be useful to the full spectrum of persons concerned with the tropics around the world—government officials, scientists, students of ecology, and others. TIE is especially pleased to thank the numerous scientists and administrators who participated in the workshop and who contributed to this report.

Arthur D. Hasler, Director, TIE

PREFACE

Ordinarily, research proceeds on a very broad front with individual investigators pursuing the questions they consider most significant and interesting. Seldom do researchers have an opportunity to review jointly the state of their subject and suggest research objectives from a collective vantage point. The National Science Foundation (NSF) of the United States of America has provided ecologists with an unusual opportunity to review the state of neotropical ecology and to recommend research priorities.

These deliberations were designed to provide guidance to the NSF on their allocation of research support. However, NSF also wanted to make the results of the study available to agencies and scientists throughout the Americas. Therefore, this research review deals with ecology in the American tropics in as broad a perspective as possible in order to be useful to a variety of scientists and decision makers in many countries and of varied political persuasions. This report is published in Spanish and English and is available at a reasonable cost to all interested persons in the Americas and elsewhere.

The TIE study is based on correspondence with more than 2,500 persons interested in tropical ecology worldwide. More than 400 persons contributed papers, advice, and ideas, and more than 100 persons actively met to discuss and write about the research needs for tropical ecology. The writers and editors of the report are acknowledged in each section. Those many other unidentified contributors we thank collectively—their support has been exceedingly helpful throughout the study. We also thank our hosts at the Departamento de Biología, Universidad de Antioquia and the Jardin Botánico for their assistance at the planning meeting in Medellin, Colombia, and the staff of the Instituto Interamericano de Ciencias Agricolas (IICA-CTEI) for their help at the Workshop in Turrialba, Costa Rica.

Finally we thank the workshop staff, Douglas Pool, in particular, and the headquarters staff at the University of Georgia Institute of Ecology, especially Silvia Halfon, Margaret Shedd, Deborah Duncan, Jimmie Sue Wagner, and Nancy Bunker, for their service to the project. We also thank Peggy Farnworth for proofreading the manuscript.

Frank Golley
Edward Farnworth

(Athens, 1974)

SUMMARY

The objective of the Tropical Ecology Research report is to review the state of knowledge in tropical ecology and to recommend profitable areas for research. The report is divided into six sections that deal with populations, communities, ecological succession, impact of technology on environment (excluding urban and manufacturing questions), regional impacts of man, and mechanisms to support research. The emphasis is on fundamental theory or patterns characteristic of neotropical populations, communities, and landscapes and on how man interacts with these units. Gaps in knowledge and research questions are identified. It is intended that the report be useful to the National Science Foundation (the agency supporting this project), and governments and scientists throughout tropical America and other tropical areas.

It was concluded that the challenge of tropical ecology is to discover ways of disturbing the natural tropical ecosystem for the benefit of man, while not jeopardizing the richest source of biological information on the earth. Population studies that consider population dynamics, genetics, environmental physiology, and natural selection are basic to understanding these dynamics of natural and man-dominated tropical systems. Specific recommendations are directed toward a sustained yield of products useful to man without degradation of productivity, richness, or long-term stability:

1. Comparative studies of the morphology, physiology, behavior, and population structure of temperate and tropical species and of paleotropical and neotropical species.
2. Comparative studies of time and energy budgets of tropical individuals and populations, assessing the position of a species in a community by examining the rates and strategies of energy acquisition and expenditure.
3. Evaluation of the impact of exotic species in order to develop better preventative and management techniques for these species.
4. Floristic and systematic studies emphasizing life history and ecological data on important ecological research sites are needed to provide background information for ecologists.

5. In planning the establishment of national parks and preserves, consideration should be given to the predictions of island biogeography theory and to results obtained in preserves already established, since extinction rates tend to increase as areas decreases.

6. The creation of greater awareness of the threat of pests to the agricultural- and forest-development programs and human populations.

7. Determination of the high-priority pest and disease problems on a regional basis and then to develop around them truly integrated pest-management research efforts utilizing personnel and financial resources from the collaborating countries and international sources. A systems approach will be essential to integrate methodologies and different disciplinary inputs into a workable, rational system of pest management at the agroecosystem level.

8. Characterization of the factors responsible for the natural population regulation of principal insect pests.

9. Epidemiological studies conducted for the more important disease agents. Also, life-system studies of competitors, hyperparasites (if any), antagonists, conditions influencing inoculum production and dispersal, host defense mechanisms, and environmental conditions critical to the disease process are needed.

10. The "center of diversity" of major crop plants should be worked intensively in search of usable genes for pest resistance. This approach should serve as the principal element in the integrated control of crop pests and possibly those of animals.

11. New breeding schemes must be devised to combine the high-yielding, uniform-product advantages of the "monoculture" system with animal and crop-breeding schemes that will endow greater genetic diversity and consequently greater agroecosystem stability.

12. Methods should be investigated for management of weed populations that would minimize the use of herbicides. Very little is known about the long-term effects of a heavy herbicide burden in the tropical ecosystem and on traditional agriculture.

Ecological research on ecosystem structure and function must be conceived, executed, and interpreted in the context of the needs and realities of society in the tropics at the same time that it must also be responsive to the further development of ecological theory. The concept of the ecological system is a useful unifying concept for the design of research questions relevant to both ecological theory and society's needs, because the needs of society are related to the forces acting upon it as part of a system. The research questions that emerge from this approach are as follows:

1. The hypothesis that process rates (e.g., productivity, decomposition, mineral cycling, succession, etc.) appear higher in the tropics than in analogous temperate regions needs to be tested in a

in a variety of tropical environments.

2. Rates of productivity, decomposition, and cycling must be known to assess, for example, potential organic yields, capacities for waste processing, and water storage alternatives. Rates of succession are requisite for developing management practices relating to ecosystem recovery silting of water bodies and devising new kinds of yield systems.

3. Both short- and long-term cyclic phenomena must be incorporated into ecosystem study.

4. External and internal factors that control ecosystem structure and function must be investigated in order to understand management of ecosystems. Natural systems provide free work and with proper management reduce maintenance costs.

5. Studies of data on the dynamics of the important nutrient-element cycles that occur during clearing and forest destruction, during cropping, and during abandonment and forest recovery are needed.

6. Determination of the optimal rates at which ecosystems operate is of primary importance to ecological theory. We also must know how much stress an ecosystem can tolerate without losing its capacity to recover.

7. Procedures for tropical research that will facilitate the development and use of the necessary interpretive framework are recommended. These procedural steps include research design in the context of the next highest level of biological organization, development of integrating mechanisms to summarize, interpret, and present the data, include energy network diagrams and models, and consistent reporting and the use of comparable techniques.

8. More data are needed to test the hypothesis that diversity and stability (or better, resistance to perturbation) are directly related.

9. Increased study of regional coupling of ecosystems is necessary since coupling involves the synthesis of theory at a level of organization that is of practical importance to the needs of society. Remote sensing, delineation, and sampling of watershed units, and techniques of modeling to plan research and manipulate data are needed.

10. Research sites should be selected not by the present distribution of stations but by the data necessary to test general principles of ecosystem structure and function, which also are applicable to the needs of man in the tropics.

Secondary succession has been studied mainly in temperate areas and knowledge of the tropical process is derived from studies in only a few localities. Many ecological studies including secondary vegetation restrict themselves to lists of the species found in secondary communities. Whereas these data are useful, few generalizations can be derived from them. Recommendations for research include the following:

1. Establishment of permanent centers (which may be existing stations) to study succession over the long time periods.
2. Establishment of replicated plots for studying short- and long-term effects of disturbance.
3. Regional studies of secondary communities should be undertaken with preferences given to areas where past history is accurately known.
4. "Ecological life histories" (autecological population studies) of organisms prominent in and/or having a conspicuous effect on disturbed habitats are needed. A given ecological life history should extend across as wide a range of habitats as possible and should incorporate some reassessment of the taxonomy of the organism.
5. Experiments in adding and subtracting specific organisms or groups of organisms (e.g., all ants, all insects, all vines) from many of the types of communities are needed.
6. Taxonomic treatments focused on prominent groups in badly disturbed habitats should be encouraged. These should proceed hand-in-hand with the development of easily accessible data-retrieval systems for all the information available (not only taxonomic) concerning particular organisms.
7. Knowledge of soil changes occurring in the course of secondary succession is needed.
8. Detailed publication of data from the studies mentioned above, in journals or other publications of wide circulation (not only data banks) is needed. It is imperative that these receive widespread translation into Spanish, English, and Portuguese.

Human use of the land can be grouped into such categories as agriculture, grazing, forestry, industry, urban development, and transportation. Each of these activities uses the environment in a special way, requiring certain inputs and resulting in specific outputs. Thus, the problem of analysis of the impact of man not only concerns the maintenance of a given system, but also the inputs and outputs of the system, which link it to all other man-dominated or natural ecosystems. Recommendations for research are as follows:

1. The environmental controls of plant and animal agricultural production in the tropics should be studied.
2. Studies of traditional crops and wild plants and animals as new food sources are essential.
3. Traditional agricultural systems, multiple cropping systems, integrated agricultural operations using plants and animals should all

be investigated utilizing modern techniques of system analysis where costs and benefits from an environmental, economic, and social point of view are considered.

4. The value of faunal and floral resources as sources of food, sport, recreation, and genetic material should be assessed.

5. Development of a sound ecological base for tropical forest management is needed.

6. The impact of reservoirs on river basins, aquatic flora and fauna, man living in river basins, disease organisms, and soil erosion should be investigated.

7. Standard methods of obtaining and reporting climatological data are needed in tropical areas since these data are fundamental to understanding and predicting processes in all ecological systems.

8. Finally, research is needed on the impact of transportation networks, mining, tourism, and other land use patterns in the tropical environment and the interaction of these activities with other activities and processes. Urban and industrial problems merit special attention.

Land, water, and atmosphere are intimately linked through complex biological and geophysical processes without respect to national political boundaries or even to continental limits. "Action at a distance" is indeed an accurate metaphor to describe the interregional and international impacts of one major ecosystem upon another. Better understanding of regional ecosystem interactions can be of substantial help in planning to maximize the contribution of natural resources to human welfare. Recommendations for research are the following:

1. Expanded studies of the biogeochemistry of major tropical drainage basins are required to establish the initial physical, chemical, and biotic characteristics of their streams and associated estuaries.

2. The relative roles of various energy sources such as mangroves, algae, and river-borne detritus in tropical estuaries should be established with particular emphasis on the effect of perturbation of these sources on food webs and fish yields. Detailed effects of alterations in fluvial regimes on the productivity of rivers, lakes, marshes, deltas, and estuaries should be determined.

3. Field studies should be undertaken to determine the quantitative importance of marshes, other wetlands and running water ecosystems for production and support of pelagic fishes and of waterfowl and other birds that migrate between continents and as habitat for resident birds and mammals.

4. Additional research should be directed toward more fully establishing the acute and chronic limits of aquatic and estuarine organisms toward pollutants, temperature, and wide fluctuations in

ambient conditions.

5. The influence of thermal and mixing regimes characteristic of tropical lakes and reservoirs on the circulation and biological effects of pollutants should be studied.

6. Before major modifications of shoreline physical conditions or of river flow regimes are undertaken, careful study should be given to their probable effect on longshore current patterns, littoral drift, and sediment accumulation or erosion.

7. An intergovernmental committee should be established to explore the desirability, feasibility, and structure of a monitoring program designed to determine the sources and distribution of organochlorine insecticides, PCBs, and selected heavy metals.

8. Each coastal nation should undertake a program to classify its estuarine resources on the basis of sound management and ecological considerations.

9. Widespread human diseases should be studied as regional ecological systems in which land-water interactions are paramount and in which social, economic, demographic, political, agricultural, industrial, chemical, geological, and climatic factors are considered.

10. An assessment of potential ecological consequences, based on explicit consideration of local climate, plants and animals, land use, patterns of human activity in the area of application should be made before initiating programs of industrialization or weather modification.

11. Field research and simulation modeling are needed to determine more accurately the annual and seasonal evaporation and transpiration rates from extensive areas of various types of vegetation, natural and man-modified, in tropical lowlands. Improved estimates of production rates, size distribution, and mean residence times of particles entering the atmosphere from burning of agricultural and forest wastes, including site preparation in shifting cultivation, should be made. Also, regular monitoring of the total biomass, and hence carbon storage of tropical forests, grasslands, and agricultural crops, is needed.

12. Modeling and prediction of ecosystem-atmosphere interactions will impose requirements for both climatic and ecological data on a greatly expanded scale. Therefore, cooperation should be improved between ecologists and atmospheric scientists concerned with global climatic measurements. The designers of metereological and climatological satellites should be encouraged to accelerate the development of methods for improved acquisition and retrieval of ecologically important climatic data by remote-sensing methods.

Tropical countries need ecological research since it is probably not possible to directly transfer information from temperate regions to the tropics. Ecological research also underlies and directly supports efforts to expand food production and industry, control disease and pests, and effectively plan the optimum use of the tropical lands and waters. Effective research requires an adequate institutional base, trained workers, and integration into a communication network so that research done elsewhere in the tropics can be applied to local problems. Finally, research has to be linked to societies' needs so that results can be effectively applied. To accomplish these needs it is recommended that:

1. Research groups be organized so that basic and applied ecologists coordinate their work. These groups should be developed at different tropical research sites, and should have an international structure because of scarcity of human resources. Two not mutually exclusive alternatives are (a) an independent coordinating office connected to an intergovernmental organization with the task of encouraging and supporting research programs and (b) the creation of an inter-American institute of tropical ecology which could serve as a training, coordinating, and information collecting structure.

2. Research projects funded or organized by nonresident scientists should incorporate local colleagues or advanced students. Funding agencies should insist that local cooperators be identified and assure that they will participate in research programs in a meaningful way. Foreign research groups should reinforce local institutions by giving advice to research programs and, in every case, information obtained through their studies should be supplied to the host institutions either as reprints of papers published elsewhere or in the form of articles published in a local journal.

3. Those few people active in ecological research in the tropics should have the opportunity of further training by means of intensive short courses in general and tropical ecology. Complementing these people should be graduates in the fields of biology, agronomy, forestry, veterinary science, and other sciences who study ecology at the postgraduate level at those temperate-zone universities that allow the student to do the thesis on a tropical problem in his home country.

4. A unified curriculum in ecology for students of tropical American countries is recommended. This curriculum should be offered in a few centers of university-level education in tropical America, with the objective of training professional ecologists.

5. In order to develop a useful and efficient communication mechanism between researchers in tropical ecology, it is suggested that the international coordinating unit previously proposed include an information center utilizing modern data-processing equipment and techniques.

6. In the interim before formation of a coordinating center and

data banks, it is desirable to have a quarterly newsletter on research in progress and recently completed (prior to publication) in the tropics.

7. The greatest possible effort should be made to reach the widest audience, including decision makers in relation to basic and applied ecology, using all possible audiovisual aids. This process would be facilitated by short courses on science and ecology for the press and audiovisual reporters.

8. Frequent international meetings should be held in tropical regions and courses dealing with ecological problems should be implemented.

TROPICAL ECOLOGY WORKSHOP PARTICIPANTS

Team 1: Tropical Population Ecology

Gordon Orians, Team Leader
Department of Zoology
University of Washington
Seattle, Washington

J. Larry Apple
Department of Plant Pathology
North Carolina State University
Raleigh, North Carolina

Ronald Billings
College of Forest Resources
University of Washington
Seattle, Washington

Louis Fournier
Departamento de Biología
Universidad de Costa Rica
San José, Costa Rica

Larry Gilbert
Department of Zoology
University of Texas
Austin, Texas

Brian McNab
Department of Zoology
University of Florida
Gainesville, Florida

José Sarukhán
Departamento Botánica, Instituto Biología
Universidad Nacional Autonoma de Mexico
Aptd. 70-233, Mexico 20, D.F.

Neal Smith
Smithsonian Tropical Research Institute
P. O. Box 2072
Balboa, Canal Zone, Panama

Gary Stiles
Departamento de Biología
Universidad de Costa Rica
San José, Costa Rica

Thomas Yuill
Department of Veterinary Science
University of Wisconsin
Madison, Wisconsin

Contributor:
William Paddock
1323 28th Street N.W.
Washington, D.C.

Team 2: Tropical Ecosystem Structure and Function

Ariel Lugo, Team Leader
Department of Natural Resources
P. O. Box 11488
San Juan, Puerto Rico

Mark Brinson
Department of Biology
East Carolina University
Greenville, North Carolina

Maximo Cerame Vivas
Institute of Marine Sciences
University of Puerto Rico
Mayaguez, Puerto Rico

Clayton Gist
Ecology Center
Utah State University
Logan, Utah

Robert Inger
Field Museum of Natural History
Roosevelt Road at Lake Shore Dr.
Chicago, Illinois

Carl Jordan
Radiological Physics Division
Argonne National Laboratory
Argonne, Illinois

Helmut Lieth
Department of Botany
University of North Carolina
Chapel Hill, North Carolina

William Milstead
Department of Biology
University of Missouri
Kansas City, Missouri

Peter Murphy
Department of Plant Pathology and Botany
Michigan State University
East Lansing, Michigan

Nick Smythe
Smithsonian Tropical Research Institute
P. O. Box 2072
Balboa, Canal Zone, Panama

Samuel Snedaker
School of Forest Resources and Management
University of Florida
Gainesville, Florida

Contributors:

William Lewis
Savannah River Ecology Lab, Drawer E
Aiken, South Carolina

Robert Johannes
Department of Zoology
University of Georgia
Athens, Georgia

Team 3: Recovery of Tropical Ecosystems

Arturo Gómez-Pompa, Team Leader
Apartado Postal 70-268
Mexico 20 D.F.
Mexico

Ana Luisa Anaya
Departamento Botánica, Instituto Bíologia
Universidad Nacional Autonoma de México
Apartado 70-233, México 20 D.F.

Frank B. Golley
Institute of Ecology
University of Georgia
Athens, Georgia

Gary Hartshorn
Organization for Tropical Studies
Apartado 16, Ciudad Universitaria
San José, Costa Rica, C.A.

Daniel Janzen
Department of Zoology
University of Michigan
Ann Arbor, Michigan

Martin Kellman
Department of Geography
Simon Fraser University
Burnaby 2, B.C. Canada

Lorin I. Nevling
Arnold Arboretum and
Gray Herbarium
22 Divinity Avenue
Cambridge, Massachussets

Javier Peñalosa
Biological Laboratories
Harvard University
Cambridge, Massachussets

Paul Richards
School of Plant Biology
University College of North Wales
Bangor, North Wales

Carlos Vázquez
Departamento Botánica,
Instituto Bíologia
Universidad Nacional
Autonoma de México
Apartado 70-233
México 20 D.F.

Paul Zinke
Department of Forestry
University of California at
Berkeley
Berkeley, California

Contributor:
Sergio Guevara
Departamento Botánica,
Instituto Bíologia
Universidad Nacional
Autonoma de México
Apartado 70-233
México 20 D.F.

Team 4: Ecological/Technological Interactions

Charles Bennett, Team Leader
Department of Geography
University of California
Los Angeles, California

Gerardo Budowski
I.U.C.N.
1110 Morges
Switzerland

Howard Daugherty
Faculty of Environmental
Studies
York University
Toronto, Ontario, Canada

Lawrence Harris
School of Forest Resources
University of Florida
Gainesville, Florida

John Milton
Woodrow Wilson Center
Smithsonian Institution
Washington, D.C.

Hugh Popenoe
Center for Tropical
Agriculture
University of Florida
Gainesville, Florida

Nigel Smith
Department of Geography
University of California at Berkeley
Berkeley, California

Victor Urrutia
Center for Tropical Agriculture
University of Florida
Gainesville, Florida

Contributor:
Enrique Beltrán
Instituto Mexicano de Recursos Naturales Renovables,
Dr. Vertiz 724
México 12, D.F.

TEAM 5: Impacts of Regional Changes on Climates and Oceans

Charles Cooper, Team Leader
Center for Regional Environmental Studies
San Diego State College
San Diego, California

J. J. DePetris
Instituto Nacional de Limnologia
José Marcia 1933/43
Santo Tomé, Sta. Fé, Argentina

James Ehleringer
Department of Biology
California State University
San Diego, California

Richard Fisher
Faculty of Forestry
University of Toronto
Toronto, Ontario, Canada

Stuart Hurlbert
Department of Biology
California State University
San Diego, California

Stephen Schneider
National Center for Atmospheric Research
Boulder, Colorado

Jay Zieman
Environmental Sciences
University of Virginia
Charlottesville, Virginia

Team 6: Mechanisms to Support and Encourage Research and Education in Tropical Ecology

Luis Montoya, Team Leader
IICA - TROPICOS
Caixa Postal 917
Belém, Pará, Brasil

Waldemar Albertin
Departamento de Ciencias Forestales Tropical
IICA-CTEI, Aptdo. 74
Turrialba, Costa Rica

Paulo de Tarso Alvim
Centro de Pesquisas do Cacau
Caixa Postal 7
Itabuna, Bahia, Brasil

Ana Luisa Anaya
Departamento Botánica, Instituto Biología
Univ. Nac. Autonoma de México
Aptdo. 70-233, México 20, D.F.

Rufo Bazán
Departamento de Cultivos y
Suelos Tropicales
IICA-CTEI
Turrialba, Costa Rica

Douglas H. Boucher
Subdirección de Parques
Nacionales
Ministerio de Agricultura
y Ganaderia
Apartado 10094
San José, Costa Rica

Richard S. Davidson
Batelle Memorial Institute
505 King Avenue
Columbus, Ohio

Gladys de Tazán
Facultad de Ciencias Naturales
Universidad de Guayaquil
Apartado 474
Guayaquil, Ecuador

Michael Dix
Departamento de Biología
Universidad del Valle
Apartado 82
Guatemala, Guatemala

Manuel Elgueta
Centro Tropical de Enseño
e Invest.
IICA-CTEI
Turrialba, Costa Rica

Ernesto Medina
Insto. Ven. Invest. Cient. – IVIC
Apartado Postal 1827
Caracas, Venezuela

Braulio Orejas-Miranda
Department of Scientific
Affairs-OAS
1735 Eye Street N.W.
Washington, D.C.

Roger Morales González
Departamento de Ciencias
Forestales Tropical
IICA-CTEI
Turrialba, Costa Rica

Roberto S. Ramalho
Escola Superior de Florestas
Universidad de Vicosa
Vicosa, Minas Gerais, Brasil

Miguel Revelo
Division de Supervision
y Control
Instituto Colombiano Agropec.
Apartado 7984
Bogotá, D.E., Colombia

Ramiro Rizzo Leal
Departamento de Biología
Universidad del Valle
Apartado 82
Guatemala, Guatemala

Pablo Rosero
Departamento de Ciencias
Forestales Tropical
IICA-CTEI
Turrialba, Costa Rica

Jorge Soria
Departamento de Cultivos
y Suelos Tropicales
IICA-CTEI
Turrialba, Costa Rica

Joseph Tosi
Tropical Science Center
Apartado 2959
San Jose, Costa Rica

Les Whitmore
Institute of Tropical Forestry
Box AQ
Rio Piedras, Puerto Rico

Observers

James Bethel
College of Forestry
University of Washington
Seattle, Washington

Richard Bradfield
Center of Tropical Agriculture
University of Florida
Gainesville, Florida

Adelaida Chaverri
Apartado 2844
San José, Costa Rica

Leslie Holdridge
Tropical Science Center
Apartado 2959
San José, Costa Rica

John Neuhold
Ecology Center
Utah State University
Logan, Utah

Sergio Salas
Departamento de Bíologia
Universidad de Costa Rica
San José, Costa Rica

Madja Zumer-Linder
Sec. Intern. Ecology, Sweden
Wenner Gren Center,
Sveavagen 166[8]
S-113 46 Stockholm, Sweden

Staff

Sylvia Halfon
Institute of Ecology
University of Georgia
Athens, Georgia

Douglas Pool
Forest Resources and Management
University of Florida
Gainesville, Florida

Margie Shedd
Institute of Ecology
University of Georgia
Athens, Georgia

CONTENTS

Fragile Ecosystems

Section 1

INTRODUCTION

In the last 10 years, popular and technical authors alike have stressed the finite nature of the planet Earth and the close integration of human societies and the nonhuman ecosystems upon which we depend. These concepts are not new. They can be traced in most philosophical systems and can often be identified in folk cultures. Yet the current emphasis, which has led to such events as the 1972 Stockholm UN Conference on the Human Environment, indicates a heightened sense of awareness and urgency about the long-term health of our supporting ecosystems.

Clearly, our knowledge of different aspects of the biosphere is unequal. Certain regions, such as the tropics, have been studied less intensively than the temperate regions of middle latitudes. This inequality of information is partly a function of history, but also is related to the complex nature of tropical ecosystems and the support base available to students of the tropics. Organizations such as UNESCO/MAB and IICA/Tropicos stress the need to concentrate research on the tropical portions of the biosphere, not just for the sake of knowledge, but also to enable the residents of these regions to utilize and preserve their resources for their own long-term benefit.

"Production," and "growth," and "development" are all key words that encapsulate the hope of mankind to further improve our condition by reducing the insecurity of existence and the inequality of resource distribution. There is much current debate on the extent of such development, the constraints to production, and the most effective patterns of resource allocation. There is, however, general agreement that knowledge of the land, plants, animals, and people will improve the management and use of the natural world and improve mutual interaction.

The object of this report is to review briefly the state of knowledge in tropical ecology and to recommend some promising areas for future research. Ecology is defined here as the study of interactions between man, plants, animals, and their environment. As such, ecology considers individual organisms, populations, and communities, as well as large units of landscape, such as oceans, estuaries, and forests.

Environment in this context is defined as all those physical, chemical, biological, and social factors that impinge upon an individual, population, or community.

The study of ecological systems should result in a more realistic understanding of the interactions between plants and animals, nature and man, and land and oceans, which in turn will permit realistic management of resources. This dual role of ecology will pervade our report, since most research questions are framed here in both their theoretical nature and in their applied form relative to the human condition.

The word "tropics" is rich in meaning for both the ecologist and the layman. In this report we have avoided a precise definition and have utilized the geographer's description of the tropics as that area of the Earth's surface between 23.5° North and South latitude. We have chosen this convenience since our ultimate goal is to say something of use to man in tropical America. This region is rich in a variety of habitats ranging from permanent snow and ice to lowland wet forests and deep ocean troughs, all used by man in one form or another. Our emphasis will be on tropical forests because forests cover so much of the land surface and because so many tropical ecologists are terrestrially oriented. However, we cannot and do not ignore tropical montane and alpine environments or rivers, lakes, estuaries, and oceans.

This report was prepared by a group of scientists gathered for a "workshop," a period of time during which they lived together and collectively discussed and wrote a report summarizing their interpretation of a particular problem. In this case, the workshop was preceded by an extensive period of private study and writing in preparation for the collective discussions. This report was requested by the National Science Foundation (NSF), which supports basic science in the United States of America. The NSF seeks advice from the scientific community about the form and usefulness of such support, with the "scientific community" understood in the largest, international sense. This has been done not only to gain the best advice possible, but also because the impact of such knowledge would be equally important in numerous countries outside of the United States of America.

NSF approached the new organization, The Institute of Ecology (TIE), to serve as an institutional base for the study. TIE was a particularly appropriate choice since it is international and has institutional members from Argentina to Canada. It also has no institutional bias toward the tropics.

The proposal written by TIE named Frank Golley, Robert Inger, Daniel Janzen, Ernesto Medina, William Milstead, John Milton, Harold Mooney, and Joseph Tosi responsible for the study. These principal investigators chose Golley as group chairman and hired Edward Farnworth as the project manager. They also developed a list of tropical ecologists and administrators of tropical ecology throughout the world who might contribute to the study. This list has since grown to exceed 2,500 persons. From the initial list of the principal investigators, a group of 30 persons was invited to meet at the Universidad de Antioquia, Medellin, Colombia, in July, 1972 to plan the workshop. The planning meeting resulted in a list of more than 200 research questions in tropical ecology, a larger list of persons to join the study, and the selection of IICA-CTEI, Turrialba, Costa Rica as the site for a workshop, where the report would be completed.

The conclusions of this valuable planning meeting were implemented by summarizing the research questions in six major research areas. These six form the basic structure of the report:

1. The nature of populations in tropical ecosystems.
2. The structure and functioning of tropical ecosystems.
3. Recovery of tropical ecosystems.
4. The nature of the impacts of man in tropical systems.
5. Consequences of large-scale changes in tropical landscapes.
6. Mechanisms to support and encourage research in tropical ecology.

Team leaders were chosen to organize the study of each of these areas and in turn each selected six or more scientists to join the investigation. The teams' preparatory work resulted in position papers, which were reviewed and incorporated into a draft report at the writing conference at Turrialba (March 24-31, 1973). This report was then extensively edited by Farnworth and Golley and submitted to the team leaders and principal investigators for further editing and approval. The very rapid development of this project was due to the cooperation of the scientists and their institutions as well as the excellent support staff at the University of Georgia Institute of Ecology (the office of the project), TIE central office at Madison, and IICA, Turrialba. Many contributors were active at each level of the development of the report, and approximately 400 people have participated directly in the project.

A final word is needed concerning the use of the report. The intention of the principal investigators, TIE, NSF, and the workshop participants, is that the report be directed to the theory and application of tropical ecology and therefore be above questions of local demands and social and cultural constraints. It is hoped that the report will be useful to all persons interested in tropical ecology. Specifically the report is aimed at research administrators and scientists, but it should also be useful to advanced students and to the educated layman. The report is designed to tell what is known and to indicate where further research is needed. It discusses how this research may be applied to specific problems but does not specify which problems should be addressed or which are of highest priority. These are left to the appropriate agencies, governments, or institutions. Not all areas of ecology are addressed by the report—for example, urban problems are not considered because this area was considered too broad to cover within the constraints of the time and funds available. Also, there has not been an attempt to reduce each section to a single style or to develop the report along one type of scientific approach or method. Where disagreement over theory or facts has emerged we have tried to present all viewpoints in concise form. These differences in approach suggest something of the vitality of the subject, the diversity of background of the investigators, and the complexity of the subject matter.

Overall Recommendations

1. The workshop emphasized that programs that coordinate and utilize the work of a variety of individual investigators or laboratories will be most efficient and productive. These coordinated programs must be based upon and interact with the research groups resident in a country, working with the assistance of extranational scientists and institutions. Agencies supporting tropical ecology should assure that this provision is followed.
2. Applied research in tropical ecology should be related to the research goals of resident scientists and laboratories.
3. Tropical ecological research should be reported in journals available to the nation where the research was done as well as journals of the scientists' own country. It is important that research be reported in the language of the tropical country where the research was carried out so that students and technicians have access to the results. A Latin American journal of ecology might be developed to serve as a medium of communication.
4. Tropical ecologists should support and encourage the exchange of scientists and students, research data and reports, and ideas and concepts between nations to avoid the nationalism of science, loss of efficiency and productivity, and duplication of effort.
5. Research in tropical ecology should be carried out in the context of theory and principles. There is little need for further descriptive work in a spirit of exploration and discovery. Studies should test and build on the well-developed theoretical base in population and systems ecology presently available. Specific suggestions for research are listed in each section of this report.

Section 2

TROPICAL POPULATION ECOLOGY

Gordon Orians, Team Leader

J. L. Apple
Ronald Billings
Louis Fournier
Larry Gilbert
Brian McNab
José Sarukhán
Neal Smith
Gary Stiles
Thomas Yuill

Contributor: William Paddock

I. Introduction

In this section, we will evaluate the current knowledge of population biology as a basis to determine research needs in tropical regions. Our main premise is that the understanding of natural ecosystem dynamics is gained through knowledge of population dynamics, population genetics, environmental physiology, and natural selection.

Most biologists believe that natural selection determines the characteristics of living organisms and that an explanation of an organism's attributes requires an understanding of how natural selection acts on these characteristics. Accordingly, this section asks questions about the adaptive significance of organismal traits and attempts to summarize our knowledge of how and why natural selection has produced them.

The basic unit of natural selection is the individual organism, which is born, gives rise to a varying number of offspring, and then dies. It is the adaptive differences of the individual genotypes in a population combined with the conditions in a given environment that produce the changes in gene frequencies basic to the process of evolution.

We begin, therefore, with a consideration of the action of the physical environment on the attributes of individuals. We have concentrated on those attributes of particular ecological significance since these are the ones used in understanding population and community-level phenomena.

Although the individual organism is the basic unit, the action of natural selection on an individual can be understood only in the context of the population to which it belongs. The fitness of an individual depends on the frequencies of different phenotypes or genotypes in the population because they determine genetic constitution of possible or probable offspring, patterns of predation, severity of competition, and other population attributes.

Therefore, following a treatment of the molding of the attributes of individuals by natural selection in tropical environments, we turn to a consideration of population dynamics. This section considers basic properties of populations such as birth and death rates and how they compare between tropical and extratropical regions. In addition, we treat the important but less understood area of population structure, the extent of genetic variability in space and time, the amount of movement of individuals from one habitat patch to another, and the social structure of the populations. These properties are vital to an understanding of how populations respond to disturbed environments and to judgments of the propensity for a species to invade and become troublesome in environments modified by man.

In addition to considering individuals of the same species living together in the environment, we must also examine the structure of the surrounding community because the coexisting species will include potential and actual competitors, predators, parasites, and mutualistic species. We review the status of the theories of competition, predation, parasitism, and mutualism and examine their consequences in tropical ecosystems. In particular, we examine patterns of species diversity and richness, phenological structure of communities, and patterns of energy flow and stability. Our objective in dealing with these problems here is not to provide broad surveys of patterns in actual communities since these will appear in Section 3. Instead, we point out how these community properties are influenced by underlying events at the population level and how they may be predicted from these events *and* a knowledge of the physical environment. The extent to which these community properties are predictable from population events is still uncertain and a satisfactory resolution of this problem will not be achieved for many years, but it is important to explore as fully as possible the current extent of our knowledge and to determine what implications are available for students of community-level phenomena.

All ecologists are concerned about the effects of massive human intervention in the tropics on properties of populations and communities. Therefore, we examine problems of pest and disease management in agroecosystems from the point of view of predator-prey and parasite-host interactions. Economically important species are viewed as prey for a variety of natural predators derived from the surrounding ecosystems. How extensive their depredations become depends on the way in which we disturb the communities and the genetic, social, and spatial characteristics of the populations we establish. Therefore, population ecology should be able to provide valuable insights into the expected consequences of our activities and to indicate how we might minimize the undesirable effects of our intervention by wise application of ecological principles and knowledge.

Finally, even though our knowledge is very incomplete and in many cases we cannot yet predict the result of various perturbations in tropical ecosystems, we offer our best judgments concerning the most serious gaps in our knowledge and what kinds of applied research should be most productive to help establish and maintain a stable man-ecosystem relationship over long time periods. These recommendations are tentative because they are based on incomplete knowledge and because they are the product of the deliberations of a rather small number of persons. Other groups of qualified persons would presumably have developed somewhat different sets of recommendations, although we would expect that there would be considerable overlap in the sets. Our objective has been to be as explicit as possible about the theoretical and empirical framework in which we have operated so that evaluation of our recommendations by others will be facilitated.

II. Recommendations for Research

The challenge of tropical ecology is to discover ways of disturbing the natural tropical ecosystem for the benefit of man without jeopardizing the richest source of biological information on the earth. An attempt must be made to establish area-specific carrying capacities for man and his domesticated organisms throughout the tropics (and other parts of the world) with the preservation of natural ecosystems entering into the calculations in each region. Recommendations are directed toward a sustained yield of products useful to man without degradation of the productivity, richness, and long-term viability of ecosystems. They are based on our best assessment of the status of knowledge about population structure, dynamics, and interactions and the implications of this information for various policy options confronting us. Our recommendations follow:

1. Since community structure depends upon its constituent species, many of the properties of a community derive from their characteristics but little is known of the biology of individual species in the tropics. Therefore, there should be comparative studies of the morphology, physiology, and behavior of temperate and tropical species and of paleotropical and neotropical species. It would be most fruitful to examine sets of "ecological equivalent" species, the members of which are geographically separated.

2. One of the most important ways in which a species is coupled to both its physical and biological environment is in its use of time and energy. Therefore, comparative studies of time and energy budgets of tropical individuals and populations, assessing the position of

a species in a community by examining the rates and strategies of energy acquisition and expenditure, are necessary.

3. At several places in our report, we have pointed out the paucity of knowledge concerning properties of tropical populations. Yet this kind of information is extremely important for the understanding of the structure of communities in tropical regions and for predicting the results of various kinds of interventions in tropical ecosystems. We urge that studies of the dynamics and structure of populations of a number of tropical species be encouraged. To maximize the usefulness of these studies it is important that the following criteria be considered: (a) Studies should be of sufficient intensity and duration so that marked individuals can be followed and their survival and reproductive output measured. (b) Studies should include a gradient of habitat types of proper size, vegetation condition, and known history of disturbance so that the behavior of the populations can be examined under several conditions. Otherwise, extrapolation from single studies may be very misleading. (c) Populations should be selected to include species representative of different taxonomic groups and different successional stages and habitats. If this is not done we may find that most studies pertain to a small subset of total species and our predictive powers will be accordingly restricted. (d) Once basic data have been obtained, populations should be manipulated so that the proximate responses to these perturbations can be followed and analyzed. Such analyses are of particular use in understanding and anticipating the responses of populations to perturbations associated with increases in human populations in those environments.

We believe that a broad range of populations should be studied but point out that plant-herbivore interactions are of great importance because herbivores, as consumers of primary productivity, are able to exert powerful influences on the structure and functioning of ecosystems. In addition, many of the most important economic problems in disturbed tropical ecosystems relate to herbivore infestations of crops. The economic implications of an improved understanding of these interactions are strong. Similarly, studies in the neglected area of seed population dynamics are needed.

4. It is commonly observed that exotic species have important effects on native species and on the dynamics of the communities into which they are introduced. Human activities disturb ecosystems and increase invasion rates of exotic species at an alarming rate. An improved understanding of the reasons for the impact of exotic species is needed to help develop better preventative and management techniques for these species.

5. If Raven et al. (1971) are correct in their estimates of the rates of taxonomic progress and the amount of work yet to be done, alpha-taxonomy on haphazardly chosen groups may yield a low return per unit investment. On the other hand, ecologists are often frustrated

by the lack of systematic information as background to any thorough ecological study. It is important that identification of taxa in ecological studies not be sloppy (big red ant #1, little yellow ant #2, etc.). Moreover, a symbiotic relationship between ecology and systematics would help develop systematics and taxonomy. Floristic and systematic studies emphasizing life history and ecological data on important ecological research sites are needed to provide background information for ecologists.

6. In planning the establishment of national parks and preserves, consideration should be given to the predictions of island biogeography theory and to results obtained in preserves already established (see Diamond, 1973). In a relatively homogeneous area, one large park would be preferable to several smaller ones, unless these are connected by "corridors" of sufficient width. Wide roads through such areas ought to be avoided because they effectively create "islands."

7. The creation of greater awareness of the threat of pests to the agricultural and forests development programs and human populations of the neotropical nations is a major need. The point that new technology cannot be introduced into these countries without appropriate attention to pest management practices must be made in a convincing manner through symposia, workshops, and other communication devices.

8. It is necessary first to determine the high priority pest and disease problems on a regional basis and then to develop around them truly integrated pest-management research efforts utilizing personnel and financial resources from the collaborating countries and international sources. In the course of establishing priorities, studies must be undertaken to quantify the economic losses attributable to pests and diseases in the tropics. This is essential not only to a broader appreciation of the magnitude of the problems but also to the establishment of priorities for both indigenous and international programs. A systems approach will be essential to integrate methodologies and different disciplinary inputs into a workable, rational system of pest management at the agroecosystem level (Watt, 1970). This may require a degree of sophistication inconsistent with the personnel capabilities of some programs in neotropical countries, but international agencies or programs could and should provide this overview synthesis.

9. Since many of the neotropical agricultural and forest ecosystems remain essentially free of the disrupting influence of pesticides, fundamental studies should be initiated to characterize the factors responsible for the natural population regulation of principal insect pests in these environments, e.g, predators, parasites, competitors, host-plant phenology, physical environment, and life histories of the insects. In brief, complete "life system" analyses on the economically important insect pests of regional significance are needed.

10. Epidemiological studies must be conducted for the more important disease agents to accomplish the same objectives as in recommendation 9 above. This assumes that etiology of the important disease complexes is known, which definitely is not the case in some instances. Our concept of the epidemiological studies of a disease complex would include the same attributes as the life-system approach for insects, viz., competitors, hyperparasites (if any), antagonists, conditions influencing inoculum production and dispersal, host defense mechanisms, and environmental conditions critical to the disease process.

11. Although great effort has been exerted toward improving crops and animals for the tropics, selection pressures have been biased in favor of yield performance, and in most cases pest resistance has either been a second-order selection criterion or ignored. Very little selection has been imposed for resistance to insect pests. The "centers of diversity" of major crop plants (Harlan, 1961) should be worked intensively in search of usable genes for pest resistance. This approach should serve as the principal element in the integrated control of crop pests and possibly those of animals.

12. New breeding schemes must be devised to combine the high-yielding, uniform-product advantages of the "monoculture" system with animal and crop-breeding schemes that will endow greater genetic diversity and consequently greater agroecosystem stability. In most cases we would expect enhanced protection in such systems against devastating pest attack and against the consequences of new and more highly destructive biotypes and races of pests that may originate in pest populations.

13. Methods should be investigated for management of weed populations that would minimize the use of herbicides (Wilson, 1969; Zettler and Freeman, 1972). Very little is known about the long-term effects of a heavy herbicide burden in the tropical ecosystem. Traditional agriculture, or its evolutionary products, cannot afford the use of herbicides on a continual basis. Research on weed control aimed at minimizing herbicidal use are few, but the approach is attracting increasing attention.

III. Action of the Physical Environment on Organism Attributes

The physical environment places constraints on the resident plants and animals. The nature of these constraints varies greatly from one tropical environment to another, yet it is generally agreed that tropical environments are physically mild, whereas temperate environments are physically harsh. An environment may be called harsh for a particular organism if it requires, for example, a modification in the energy budget of the organism compared to what is expected of it in other environments to compensate for the physical conditions it faces. Obviously, such a response is graded. Thus, if one or more environmental parameters are sufficiently harsh for an organism the reallocation required in its energy budget may make the environment uninhabitable. The nature of these interactions and adjustments is the subject of this section.

A. Climatic Patterns

Within lowland tropical regions seasonal fluctuations in temperature are minimal and are usually considered to be well within the range of tolerance of most organisms, even though the mean temperatures range from those of a lowland rain forest to those of the highland paramo. Daily variations in temperature may be important, especially in plants or animals that are exposed to intense solar radiation. Many of the plants that encounter an intense radiation load are pioneer species but canopy species also face similar environments. In tropical regions, variations in the seasonal patterns of rainfall are of major significance and vary from those found in humid rain forests to those of tropical deserts. The wet-dry season often replaces temperature as a major cue to timing biological processes in the tropics. However, another important component of tropical seasonality is its low degree of predictability. A dry season, for example, may occur every year, but the time of its onset and its duration may be highly variable (see Fig. 2.1 for an example from Caracas, Venezuela). The variance in these factors will exert a powerful influence on the evolution of population characteristics (Levins, 1968).

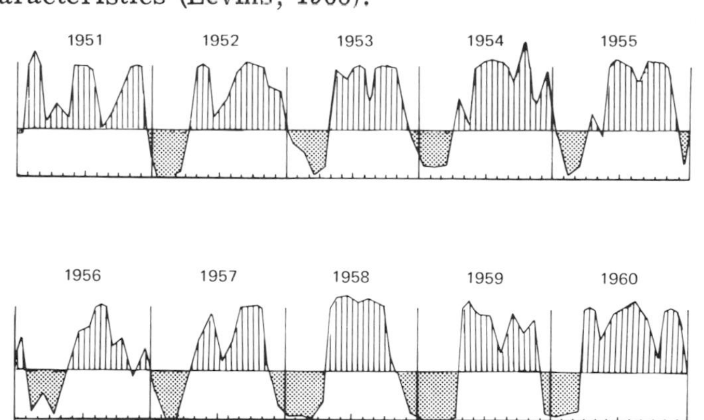

FIG. 2.1. Climate for Caracas, Venezuela. Redrawn from Walter and Medina, 1971. Dotted areas indicate dry season, lined areas indicate wet season. Note the variation in these seasons year after year.

To take a simple case, suppose that a bird requires an energy reserve to carry it through a dry season of low food availability. If the onset and length of the dry season were completely predictable, there would be a fixed optimal energy reserve required at a specific time. However, if there is variability in the length and onset of the dry season, there is no single optimal strategy. If there is no loss in fitness associated with accumulating excess energy, then there is no adaptive problem. Assuming that energy is freely available, natural selection simply favors the accumulation of enough reserve to survive through the longest unfavorable period that ever occurs. However, if being encumbered with additional energy storage reduces fitness because of a factor such as increased metabolic drain as a result of moving the larger organism around or increased vulnerability to predation because of reduced mobility, then there is no simple solution. During years of shorter dry seasons, those individuals that accumulate less reserve will be more fit, whereas in years of longer dry seasons, those individuals with a larger reserve will be more fit. The result is that the amount of variance (both genetic and phenotypic) in the population is a direct function of the amount of variance in the unfavorable period.

The theory relevant to this problem is relatively straightforward, but as yet there have been few attempts to relate the genetic and phenotypic variance in tropical species to survival in variable conditions. An important exception is the analysis of phenotypic variance in Puerto Rican *Drosophila* by Levins (1969). This theory is of vital importance in the management of pests of tropical agriculture, because control measures produce different effects in populations of low and high genetic variance and man's intervention becomes a powerful selective agent, which rapidly changes the genetic properties of the species. The rapid evolution of resistance to DDT and other chemical pesticides is the most important case in point. Insects would be the most suitable and important organisms on which to focus studies of this phenomenon because of their economic importance and short generation times, especially in tropical regions.

Finally, climates in the tropics have not been constant over recent geological time. The available evidence from deep-sea sediments, pollen cores, and other traces of past plant communities, although not totally consistent, suggests reductions in mean annual temperature in the tropics as much as 5°C within the Pleistocene (Hammen and González, 1960a,b; Moreau, 1963; Sarukhán, 1968). Considerable evidence from patterns of bird distribution also suggests major vegetational shifts in the tropics (Haffer, 1969; Vuilleumier, 1971).

The great diversity of tropical systems, therefore, is not necessarily due to their long evolutionary history and lack of recent disturbance, as many biologists have thought, but may be due to more immediate conditions. Equilibria in tropical environments may be expected in short time spans following climatic perturbations just as equilibria can be established on islands in remarkably short time spans (Simberloff and Wilson, 1969, 1970). Further analyses of recent climatic changes in the tropics and their influence on patterns of species distributions and abundances are needed. We are operating primarily on the basis of unexamined assumptions.

B. Morphological and Physiological Responses to Climate

Presumably, many of the morphological and physiological characteristics of plants and animals are responsive to the climatic conditions under which they live, although little is known of this interaction in the tropics. Most of what we know about the physiology of animals and plants derives from work on temperate species, simply because most biologists live in temperate regions. Thus, the warm-cold seasonality and the use of photoperiodicity to time annual events are basic to our concept of organisms.

One of the clearest expressions of a differential response to the physical parameters in the environment is found in the morphology of plants and animals. The theory of the adaptive significance of plant shapes has been most recently reviewed by Horn (1971), who explores the adaptive response of plants to the characteristics of light. A surprising amount of variation in plant shapes can be explained by this one variable, but the tropics exhibit some interesting differences from the temperate zone.

A conspicuous difference is the presence of monolayered trees in early successional communities, present only in climax communities in the temperate zone. The explanation for this difference may lie in the luxuriance of early successional growth in the lowland humid tropics, which makes it highly unlikely that lower layers of leaves would receive enough light for it to be an energetic asset to the plant.

The shape of leaves also reflects an adaptive response to climate. The general pattern of tropical leaves is fairly well known. Irregularly shaped leaves are rare and limited to early successional stages; virtually all leaves are regular in outline; leaves of tropical montane trees are smooth-margined at mean annual temperatures equivalent to temperate latitudes where nearly all the leaves have irregular outlines; leaf size decreases with altitude. These difference in leaf shape may be explained in many ways (see Horn, 1971). Irregular leaves have lower equilibrial temperatures when exposed to the sun. Large regular leaves in plants found in early successional stages tend to tear and thereby increase convective heat loss, lower leaf temperature, reduce water loss, and increase the rate of photosynthesis. Trees that dominate the canopy often show a diversity in leaf size and shape, or in number of stomata, depending upon whether the leaves are exposed to the sun or are normally shaded.

There may be other factors influencing the shape and size of leaves. Thus, Jordan (1971) points out that tropical forest trees produce as much dry weight of leaves each year as of wood, whereas temperate forests produce two to six times as much wood as leaves. The energy content (calories per gram of dry weight of leaves) is lower in tropical than in temperate forests. Clearly, an analysis of plant strategies from the viewpoint of examining the pattern of natural selection on the shapes and energy commitments of the plants is needed to supplement examinations of energetics at the community level.

Very little is known about the influence of the physical environment upon the physiology of tropical animals. Comparative analysis of the energetics of temperate and tropical endotherms suggests that the level and stability of body temperature is determined by the rate of heat production, insulation, and body size (McNab, 1970). These parameters, in turn, are fixed by their interaction (McNab, 1970), climatic conditions (Scholander, et al., 1953), and type of food used (McNab, 1969; Yarbrough, 1971). Because of the great reduction in the physical restraints placed on endotherms in at least the lowland tropics, a much greater range of acceptable rates of heat production exists, explaining the fact that nearly all "cool" species of mammals (such as sloths, anteaters, and armadillos) are found in the tropics.

Comparative studies on the environmental physiology of indigenous human populations in the tropics also are greatly needed before their acculturation or extinction. Various studies on the temperature regulation in African bushman and Australian aborigines have shown that these humans are more tolerant to cold exposure than are Europeans (Hammel, 1964). However, there are few data on the energetics of peoples native to lowland rain forests.

Mild tropical climates also influence ectotherms. For example, a much smaller fraction of terrestrial ectotherms in the lowland tropics is heliothermic than is the case for temperate ectotherms (Inger, 1959; Ruibal, 1961). Furthermore, the daily activity cycle in many nominally diurnal ectotherms (such as dragonflies, butterflies, and beetles) has been extended into early morning hours, limited only by available light. In most temperate environments a rise in ambient temperature or exposure to the sun is required for these insects to obtain the required flight temperatures.

Life in aquatic environments in the tropics, however, may be harsher than that usually encountered in temperate environments. Fresh waters in the tropics generally are warm with high CO_2 and low O_2 concentrations. Such an environment makes respiratory gas exchange difficult. Thus, many fish in these waters are either obligate or facultative air breathers (Carter and Beadle, 1931a,b; Wilmer, 1934).

Organisms may also respond to seasonal variations by seasonal migration or entrance into a resting state. The escape from hostile conditions in the lowland tropics by entrance into a resting state is influenced by high environmental temperatures. Many groups of insects, such as Odonata and Orthoptera, pass the unfavorable period as adults in the tropics, retreating to smaller and smaller areas of suitable habitat. At higher latitudes these groups spend the unfavorable period in an egg stage (Corbet, 1962; Janzen, 1973a,b).

As yet there is no solid body of data on the metabolic rates of insect eggs or pupae under tropical conditions. These rates have implications for the evolution of egg size in tropical insects, the ease with which the egg stage can be used to pass unfavorable seasons, the influence of egg parasites on egg size, and the evolution of life cycles with and without resting eggs. Our ability to control insect populations by techniques oriented toward reducing survivorship during the unfavorable period may be enhanced by better understanding of the adaptive significance of these life history patterns.

The analysis of the influence of climate on the morphology and physiology of both plants and animals can be best evaluated by the study of species that apparently are ecological "equivalents." Such comparative studies will be particularly effective when species from the old-world tropics are compared to those of the neotropics since the climate, food, and natural history can be matched more readily than is possible between temperate and tropical species. For example, each of the large tropical land masses has mammals that feed almost exclusively upon ants and termites, a food habit that is rare in temperate climates. These mammals represent all three mammalian suborders (monotremes, marsupials, and placentals) and include some of the most distinctive mammals in southern continents (tamanduas in South America, echidnas in Australia, and pangolins and aardvarks in Africa). These mammals, although unrelated to each other, have low body temperatures (the levels of which can be accounted for by body size), very low rates of heat production, and poor insulation (McNab, unpublished observations). It would appear that their low rates of metabolism are related to their distinctive food habits.

Comparative studies of individual species have a practical side, too. There are strong tendencies for temperate man to export his animal husbandry as well as other technologies to tropical regions, but there is no assurance that either should be encouraged without the judicious examination of alternatives. For example, temperate cattle usually cannot maintain high meat or milk production in the tropics—that is why various hybrid stocks, usually involving Indian cattle, are now used extensively (MacFarlane and Howard, 1972).

Finally, the effect of climate on animal and plant biology can be examined by study of the tropical-temperate exchange. Is it more likely for a temperate species to invade the tropics, or vice versa? Possibly the limitation to temperate species invading the tropics is "competition," while the limitation to tropical species is the physical harshness of a temperate environment directed either at the potential invaders themselves, or at their food supplies (MacArthur, 1972).

Yet the actual situation is much more complex. For example, many temperate bats are not found in lowland Mexico, where there are many tropical bats, but few temperate bats reach southern Florida, where there are almost no tropical species. There is considerable evidence that temperate cave bats at least are limited in their southerly distribution by the necessity to hibernate. One species at its southern limits even migrates northward to spend the winter in caves that are sufficiently cold for them to hibernate. The exchange of these mammals, as well as other animals and plant populations, is a fruitful area for investigation.

IV. Population Structure and Dynamics

A. Numerical Properties of Populations

Age-specific birth- and death-rate processes determine the age structure of a population, and estimates of age-specific fecundity and survivorship allow the calculation of the mean reproductive value for any individual of some age at a certain time. The reproductive value curve for each population is an important aspect of life history information in formulating theoretical predictions and management decisions. Such demographic data are rare for tropical organisms in general and long-term studies of marked populations of plants and invertebrates are badly needed for tropical environments.

1. Plant populations

Demography and population dynamics have been classically the realm of zoologists and have constituted the backbone of studies on mechanisms of population regulation. Much of our present knowledge of theoretical population biology has been upon studies of animal demography. Plant ecology in general and tropical plant ecology in particular has contributed very little to the understanding of the regulation and dynamics of natural populations.

Few studies have provided actuarial data about plant populations. These studies have dealt primarily with herbaceous species typical of temperate grassland communities (Tamm, 1948, 1956, and analyses of his data by Harper, 1967; Canfield, 1957; Rabotnov, 1956, 1958; Sagar, 1959, 1970; Antonovics, 1966, 1972; Williams, 1970; Sarukhán, 1971, 1972, in press; Sarukhán, and Gadgil, in press; Sarukhán, and Harper, 1973). Hartshorn's (1972) study of *Pentaclethia macroba* and *Stryphnodendion excelsum* in Costa Rica is the only source of information on population dynamics of tropical plants. In this work the effect of physical and biotic factors upon the survival of plants at different stages is considered and a predictive model is proposed.

The paucity of demographic studies in the plant ecological literature could be due to several factors. The following three are perhaps the most important: (a) Plant populations are of such obvious importance in determining the biotic environment of communities that ecologists have been preoccupied with the study of their composition, structure, spatial distribution, and relation to their physical environment. (b) Certain factors that affect population size in animals act upon plant populations with very different results, viz. (i) predation of animal on animal usually means the death of the individual, whereas "predation" or rather "parasitism" by animals on plants may occur without immediately obvious demographic consequences, and (ii) migration with the resulting change in numbers in an area is a common reaction of animals confronted with hostile or unfavorable environmental conditions, whereas a high degree of plasticity is the usual response of spatially fixed plants to an unpropitious environment. (c) Plasticity may considerably deform the meaning of numbers in plant populations, since individuals can literally vary a thousandfold in their reproductive behavior (Harper and McNaughton, 1962; Harper and Clatworthy, 1963). Studies of the fate and behavior of marked individuals provide understanding of the dynamics of plant population regulation, knowledge that is basic to forest management and weed control in agroecosystems.

2. Seed populations

Seed populations have suffered from even more neglect than mature plant populations by plant ecologists. At the very descriptive level, floristic lists of seed contents in the forest soil (which are relatively common for some temperate regions) are virtually absent in the tropics, apart from a few carried out in Puerto Rico, Mexico, and Malaya (Bell, 1970; Guevara and Gómez-Pompa, 1972; Symington, 1933). It is impossible to attain a thorough understanding of population dynamics of plants if only a portion of the population (the mature plants) is considered. There is ample evidence that at least for a number of species, selective forces act more strongly on the seed and seedling populations than they do in mature plants.

Seeds are a special case because they contain a very nutritious meal neatly packaged for a potential predator. Seed predation rates are regularly very high in the tropics and are often the result of highly specialized insect seed predators, although generalized seed predators are also present. Available theory to explain patterns of seed characteristics and their evolution in response to the physical environment and herbivore pressure does not agree on all points (Janzen, 1969, 1971a; Smith, 1970) and needs to be tested further. This interaction is important since there will almost certainly be attempts to reduce loss of agricultural seeds to predators, yet we wish to avoid the environmental damage caused by some of the techniques employed in temperate localities.

Although descriptive work on adaptive morphology for seed dispersal exists (van der Pijl, 1968), studies on the demographic consequences of these adaptations are still to be done. The determination of the factors that affect dispersal distances are necessary for a better understanding of predator-seed relationships. Furthermore, comparative studies of dispersal patterns of seeds of species in primary and secondary vegetation do not exist.

We have very little information on the life span of seeds of tropical species and the factors that determine life span (Marrero, 1943; Vázquez and Gómez-Pompa, 1972). In fact, this is a subject of tropical biology filled mostly with anecdotal information. We suspect that seeds of species of primary forests are very short lived, in contrast to those of species of secondary vegetation, which possess considerable longevity. We also suspect that the seeds of pioneer and secondary vegetation species have mechanisms that allow them to switch between different types of dormancy in order to avoid germination at unsuitable periods and, at the same time, to optimize germination at times when the probability of individual survival is highest.

Many fundamental questions about adaptive characteristics of seeds await an answer not only in tropical but also in temperate areas: (a) Has selection acted to reduce the germinating fraction of "secondary species" present in the soil of primary forests (thus increasing their viability) because the probabilities of a seed of these species yielding a reproducing adult are very low under the conditions of the primary forest? (b) Does the same principle apply for very early pioneers? (c) On the other hand, has selection maximized the germinating fraction of seed of primary species, therefore avoiding dormancy that has no selective value under the stable conditions of the virgin forest? (d) How is this situation affected under the present intensive disturbance to which tropical vegetation is being subjected? A compelling theory on seed germination strategies awaits validation with field data (Cohen, 1966, 1967, 1968).

3. Insect populations

Study of the dynamics of insect populations has historically emphasized economic pest species (Clark, et al., 1967; Boer and Gradwell, 1971). Recent studies (Gilbert and Singer, 1973) indicate that this emphasis has resulted in models of insect populations that are not sufficiently general for strong statements or predictions about a randomly selected insect. This problem is compounded in the tropics where studies are few and where both species and the diversity of life history types are large.

Aside from detailed studies of such tropical insect pests as the tsetse fly in Africa, *Aedes aegypti* in Asia, and ambrosia beetles in Malayasia (Browne, 1961), adequate studies of tropical insects other than crude estimates of abundance during various seasons are not available. However, recent long-term marking studies on certain tropical butterflies provide a baseline for a discussion of possible differences between temperate and tropical insects populations as well as between disturbed versus natural sites in the tropics. Owen (1971) has studied *Acraer encedon* which as a larva feeds on a weed *Commelina* in Africa. He indicates that this species, aside from being present year around, is not unlike ecologically similar temperate species such as *Euphydryas* sp. and *Chlosyne* sp. in terms of reproductive value curves, egg sizes, larval biology, adult longevity, population size, and population stability. In striking contrast, a 2-year study of *Heliconius* in Trinidad (Ehrlich and Gilbert, 1973) revealed that marked adults live at least 6 months with no inactive period. Moreover, egg production is evenly timed as a result of the adult stage assuming much of the burden for collecting necessary resources. Assimilation of nitrogenous compounds by adult butterflies was thought nonexistent or unimportant until the discovery that *Heliconius* collects pollen, extracts amino acids, and uses these in the production of eggs (Gilbert, 1972). The plants that provide pollen do so at a regular rate throughout the year, so that even during the dry season when many adult butterfly populations decline, *Heliconius ethilla* populations remain constant. In fact, a constant population size was maintained for at least 27 generations, a striking figure for any organism. Further, a strong mutualism exists between *Heliconius* and its pollen plant *Anguria* (Gilbert, 1972). *Anguria* possess storage tissue that allows flowering to continue through the dry season thus providing the kind of biotic buffer to harsh periods which Pianka (1966) predicted existed in the tropics.

Other tropical forest insects may follow similar life history and population dynamic patterns. For instance, euglossine bees live up to 1 year (Janzen, 1971b), have seemingly constant population levels, and depend upon widely spaced plants that provide pollen and nectar in low amounts over long periods. The potential for greater adult life spans for insects in the tropics results from both resource availability and warm year-round temperatures. The greater age at which maximum reproductive value is attained means that for many tropical insects a relatively greater investment of the total energy budget in adult structures (flight muscles, brains, sense organs) can be expected since these adults must forage and compete for limited resources. Undoubtedly further studies of insect populations in the tropics will reveal further properties that can not be predicted from the current literature on insect population biology.

4. Vertebrate populations

Studies of populations of marked individual vertebrates over several years are extremely rare. Outstanding exceptions are the detailed studies of Panamanian birds by Willis (1967) and Snow's (1962) work on the black-and-white Manakin of Trinidad. Ricklefs (1969) has summarized the few other data on tropical-bird population dynamics. These studies confirm the anecdotal evidence that mortality rates on adult birds in the tropics are lower than in the temperate zone but that nest predation rates are higher. In general, reproductive rates also are lower in most tropical birds (Lack, 1954, 1968; Cody, 1966), lizards (Inger and Greenberg, 1966; Tinkle, Wilbur and Tilley, 1970), frogs (Inger and Bacon, 1968), and mammals (Fleming, 1970, 1971).

The theory of natural selection acting on reproductive rates is well advanced (Cole, 1954; Gadgil and Bossert, 1970; Lack, 1954; MacArthur, 1968; and Salisbury, 1942), but the causes of the low reproductive rates are unresolved. Theories implicating the role of food shortage and predation have been combined by Cody (1966) but critical field data are still lacking.

Indirect evidence indicates that tropical vertebrates, especially those species of mature forests, are remarkably sedentary. Gaps of unsuitable environments as small as a few hundred meters appear to be major barriers to tropical forest birds (MacArthur, 1972; Diamond, 1973), and the same may be true for mammals and reptiles. This results in highly inbred local populations, many gaps in the ranges of species, and a high incidence in the formation of local races. These patterns are quite clear and can be deduced from static data, but knowledge of population densities and population dynamics, information that requires detailed studies over longer periods of time, is meager.

B. Population Structure and Patterns

A major set of questions in population biology involves the factors that give the individuals, genotypes, or genes of populations their distribution in space and time (Ehrlich and Holm, 1962). The structure of a population is determined by the behavior of the individuals, which in turn are selected to exploit their environment in the most effective way possible. Each species and in some cases each local population of a species faces a different distribution, abundance, and predictability of resources, as well as different competitors and predators in their environment. The theory of population structure response to changing or heterogeneous environments is well developed (Levins, 1968) and allows predictions concerning the adaptive significance of dispersal and gene flow (see Gilbert and Singer, 1973) and the maintenance of genetic variability in populations. The theory of patterning of individuals in space is less well developed. Yet, knowledge of how tropical populations are structured and how they will respond to the radical changes now occurring in tropical regions is highly important from both a theoretical and practical point of view.

Spatial isolation of tree species is a common feature of some tropical forests and an explanation for such spacing has been proposed by Janzen (1970a) and Connell (1970). However, where environmental factors such as soil result in communities dominated by few arboreal species, patterns of distribution departing notoriously from hyper-dispersion may occur and the causal factors of these patterns have not been studied. Equally, the uniform distribution observed for tropical trees may not necessarily apply to the patterns of distribution of other components of the plant community, such as lower story species. Further, a change in the dispersal pattern of plant species seems to occur during the life of the individuals. Connell (1970) has observed an increase in "pattern" diversity with increasing plant age caused by biotic factors. However, information on this type of patterning is very scanty and corroboration of structural change with age is needed. A similar analysis should be made of structural changes of secondary vegetation communities.

Inbreeding seems to be a common feature for most tropical plants because of the spatial isolation of individuals and the growing evidence that local populations of animals act as pollinators of many plants. However, this field has rarely been explored. Further, the implications of human disturbance on the gene flow of species of tropical primary and secondary communities should be studied.

Detailed data on population structure are available for only a few tropical insects, perhaps the most instructive of which is the study of *Heliconius ethilla* considered in the previous section. Further, Pittendrigh (1950) worked on the temporal and spatial structure of mosquitos in Trinidad. Much information on dispersal in tropical environments has been gathered for pests such as locusts and the tsetse fly in Africa. Genetic structure of *Drosophila willistoni* has been determined with respect to a few enzyme loci and inversion polymorphisms (Ayala et al., 1971; Da Cunha et al., 1950). However, local distribution and dispersal of tropical *Drosophila* has been studied only recently in Hawaii.

Tropical insects that depend upon secondary successional plants probably have greater dispersal tendencies by individuals and less local genetic differentiation. Even within a species it is known that some populations may be more fugitive than others (Gilbert and Singer, 1973). Detailed studies on the same species in different parts of its range are needed for a representative group of tropical insect species.

Social insects, although they occur at all latitudes, are concentrated in the tropics but ants and termites are strangely lacking at high altitudes (Janzen, 1973a,b). Wilson (1968, 1971) has suggested that more constant environments with less fluctuations in resource availability would result both in more specialization of castes and in relatively constant numbers of individuals in each caste. The abundance of these insects makes them among the chief predators, scavengers, and herbivores in the tropics. However, studies on the structure of social insect populations in the tropics are practically nonexistent.

A number of nonsocial tropical insects posses rather complicated social behaviors such as gregarious roosting in *Heliconius* and in *Acraea* (Owen, 1971). Population studies are needed that relate resource distribution and abundance to the individual and social behavior patterns that have evolved in tropical populations since these attributes partly determine population structure.

A major selective factor on the evolution of vertebrate social organization is the nature of the resources being exploited by the animals. Most current theory suggests strong interactions between the temporal and spatial variability of the resources, the organism's ability to defend the resource, and the resulting patterns of clumping, spacing, and group sizes (Brown, 1964; Brown and Orians, 1970; Crook, 1965; Grant, 1968; McNab, 1963; Schoener, 1968). Future research will show to what extent these attributes of animals are predictable from energy budget considerations and from other factors, such as predation.

The concept of territoriality has occupied an important place in ecology since Howard (1920) first pointed out its prevalence among birds. Today there exists a wealth of empirical evidence and a substantial body of theory that relates the extent of spacing and/or clumping of animals to resource patterning, but spacing patterns among tropical species are poorly understood. There is a relative paucity of tropical population studies because of the difficulty of observing territorial patterns. However, birds may occupy their territories for the duration of their lives (Willis, 1967), hence the initial establishment of territories that can easily be observed is a much less frequent phenomenon. We know almost nothing about the dynamics of these systems or how prevalent territorial behavior is among tropical vertebrates. Since longterm studies with marked individuals will be necessary in most situations in the tropics, this research will be difficult.

The theory of the evolution of vertebrate mating systems has been explored in some detail (Downhower and Armitage, 1971; Orians, 1969b; Selander, 1965; Verner, 1964) and the remaining interesting cases among vertebrates are largely confined to the tropics. They involve the evolution of polygyny at least six independent times among tropical birds and the existence of unusual forms of mating and group courtship in a variety of fruit-eating tropical birds (Lack, 1968). Current theory does not adequately explain the evolution of these systems and their prevalence in tropical environments is puzzling.

C. Energy Budgets

An ecological view of plant and animal energetics harmonizes the apparently distinct areas of physiological ecology, resource partitioning, and community structure. The most convenient expression of this interaction is an energy budget, the time-dependent balance between energy acquistion and energy expenditure that every organism must maintain.

The rate of energy acquisition depends upon many factors; one of the most important is the rate of renewal of the energy resource. If the rate of renewal of resources is essentially instantaneous (e.g., light), organisms may exploit the environment from a fixed spot. However, even this system is subject to temporal variation under special conditions. Thus many understory plants have periods of flowering and growth following a leaf fall in a deciduous forest (Janzen, 1972). The limitation to energy acquisition under instantaneous renewal depends upon the physical characteristics of the leaves. Such factors as the size and shapes of surface areas and their ratio to the volume of the structure become important in setting limits to the energy acquisition.

If the rate of resource renewal is intermittent and limited to a level below that at which a stationary organism can harvest this resource, feeding requires movement and is likely to be sensitive to temporal patterns of resource availability. There may be periods in which feeding may be more important and other periods in which feeding may be less important. For example, an organism may have to time its harvesting of resources to obtain maximal returns for its foraging effort (e.g., hummingbirds, Hainsworth and Wolf, 1972). The periodicity may be hourly, daily, or seasonal, depending upon the particular resource. We need to know much more about the resource spectrum in the tropics, but it is clear that a greater variety of resources is available for a longer period of time in the tropics than in temperate regions.

Certain animals in the tropics have responded to a seasonal variation in food by migrating to other nearby regions where food is available. Although long distance migrations are more characteristic of high latitudes, seasonal movements correlated with wet and dry seasons are prominent in the tropics, especially in areas with severe dry seasons. Altitudinal migrations are also common among some birds, most notably hummingbirds. These movements are triggered by changes in population density and food supply and the same factors are probably involved, either in the proximate or ultimate sense, in most tropical migrations. The most famous and best documented case, that of the migratory locust in Africa (Kennedy, 1956), is unfortunately not a representative one. Among tropical birds, widespread intratropical migrants such as the yellow-green vireo (*Vireo flavoviridis*) do not breed in mature tropical wet forest but are associated with seasonal and successional forest types and may be subordinate to full-time residents with similar food habits. This presumably reflects a greater seasonal pulse of food in those environments, but no precise measurements are available.

Of perhaps more general significance is the annual influx of tremendous numbers of temperate-zone migrants into the tropics during the northern winter. The paleotropical situation has been extensively analyzed by Moreau (1952, 1966) who demonstrated that the migrants do not penetrate undisturbed wet forest but concentrate in seasonal environments. The same appears to be true in the neotropics as well (Slud, 1960), although data are not extensive. A basic unsolved question is the impact of these migrants upon the resident tropical species (Miller, 1963). This impact may be considerable since the greatest concentration is present during the dry season in the tropical areas north of the equator, the time when food supplies (such as insects) appear to be at a minimum (Janzen, 1973a,b).

We also need to know more about the patterns of exploitation of these resources for tropical species. The only organisms well known in this regard are birds, but even here our data on actual resource utilization are fragmentary and most of the patterns are inferred from other data, such as bill sizes and shapes, similarities among coexisting species, foraging heights and locations, and foraging techniques. For other organisms our data are insufficient to make even crude generalizations about diets and means of exploitation of the environment.

One interesting aspect of resource utilization is that food supplies may be partitioned by particle size, which normally is partially accomplished by body-size differentiation in the consumer (Hutchinson, 1959; Schoener, 1967; McNab, 1971a,b). Body size is also the most important single parameter in determining physiological functions (McNab, 1970). Thus, the very procedure of partitioning resources by size differentiation has immediate physiological consequences, almost imparting a physiological "structure" to a community of populations sharing a similar set of resources. This physiological differentiation, of course, would be most highly refined where there is the greatest partitioning. This generally means the lowland wet tropics. But here again we have the least physiological information.

A basic problem that must be solved by all living organisms is getting enough of the right kind of energy at the right time and at a minimal risk and cost to the forager. Optimal foraging theory is now reaching a fairly sophisticated level (Emlen, 1966; MacArthur, 1972; Schoener, 1971; Tullock, 1971), but actual field tests are almost totally lacking. A number of components need to be investigated to develop this theory to the point where it will be useful as a predictive tool. The theory says that a predator should include a prey type in its diet if and only if the energy gain per unit time spent foraging is increased by adding the item. To determine the foraging strategy it is necessary to know the expected energy gain per unit time by a predator as a function of prey type and the expected relative encounter rates of these types of prey. If the higher ranked prey are sufficiently common it never pays a predator to take lower ranked prey no matter what their abundances. The prediction then is that as prey abundances decreases the predator should enlarge its diet, and vice versa. No test of these predictions has been made on "particle feeders" in the tropics, although in hummingbirds the energetic parameters of nectar (such as concentration, extraction costs, and special patterns of availability) seem to be important (Hainsworth and Wolf, 1972).

The selection of a foraging patch is a similar problem. Patches can be ranked according to the average energy gained per unit time spent foraging in the patch. If possible, the predator should remain in the most productive patch until prey abundances have been reduced sufficiently to make the patch no better than the next best one. To accomplish this, however, the predator needs information about prey availabilities in different patches and it may be costly to acquire this knowledge. The investment to be made in prey acquisition should evolve in relation to the amount of variation in resource availability in space and time. The noncurious and often apparently lazy behaviors of many tropical organisms may reflect their evolution in very predictable environments. Also the great range of resource types and the lesser fluctuations of these resources in time may favor more restricted diets in tropical organisms. Nevertheless, actual comparative data on the diets of temperate and tropical organisms are scarce and our ability to relate them to spatial and temporal variations in resource availability is extremely poor. "Prey" consists of more than "energy," and also is important in terms of its mineral and molecular content. Both factors may also be used in the above analysis.

The rates of energy expenditure vary with the metabolic activities of plants and animals and with environmental parameters. Ambient temperature is an important variable. Most biological processes have a Q_{10} approximately equal to 2 or 3 and, as a consequence, proceed at high rates in the lowland tropics. The consequences of this are not necessarily favorable. For example, there is evidence that the high midday temperatures in the lowland tropics are detrimental to photosynthesis of cool adapted plants (Mooney and West, 1964). Holdridge (1967) has built this parameter into his system of world vegetation types by recognizing an upper limit above which there is no net energy gain to a plant. In the lowland tropics, temperatures remain very high all night, resulting in high nocturnal respiration rates.

The combination of these factors may result in a lower fraction of the total net-energy budget that is channeled to production in a plant at lowland levels than at lower montane sites where daytime temperatures are not inhibiting to maximal photosynthesis and where nighttime temperatures are sufficiently lower to make possible substantial energy savings. Evidence in support of this idea comes from the fact that the diversity (as measured by number) of species of plants (Fournier, 1969), insects (Janzen, 1973a,b), and birds (Orians, unpublished data) is greatest at low montane altitudes in Costa Rica, in spite of the fact that the energy budgets of endotherms may be appreciably higher at the montane sites because of the lower mean temperatures.

At still higher elevations in the tropics, mean annual temperatures are much lower and, contrary to the situation in temperate localities, there is no time during the year when warm weather is regular. As a result the gradients in diversity and abundance of various plant and animal species with altitude in the tropics are often very sharp. Insects are very strongly affected (Janzen, 1973a; Schoener and Janzen, 1968), and at high altitudes birds and bumblebees become the almost exclusive pollinators of plants (Cruden, 1972; Heinrich and Raven, 1972). Not surprisingly, the percentage of wind pollinated species apparently increases with altitude in the tropics.

We urgently need studies of energy budgets of plants and animals under lowland and montane tropical climates to understand community structure, community patterns of energy transfer, and patterns of species richness (Holdridge et al., 1971). Only studies on the energies of the flower-hummingbird interaction in the tropics are relatively well advanced.

V. Consequences of Population Interactions

The three fundamental kinds of interactions among organisms in ecosystems are those when (a) two organisms may overlap in their resource utilization and the harvesting by one adversely affects the resources available to the other (competition), (b) one organism may use another as its food source (predation, parasitism), and (c) two organisms may cooperate in resource acquisition or in the exchange of resources for services such as pollination or defense (mutualism). Although usually treated separately, these processes are aspects of resource harvesting and many of the same principles apply to all three.

A. Competition

Whenever species (or individuals within a species) overlap in their utilization of resources so that an increase in the population density of one species causes a decrease in the population density, and vice versa, we say that competition has occurred. In general competition is difficult to observe directly and most of the evidence for its importance is indirect (Orians and Collier, 1963). Advances in our understanding of competition and its results have been very rapid in recent years and anything written today probably will soon be out of date. Nevertheless it is essential to provide a concise summary of current concepts since they form the basis of our interpretation of processes in tropical ecosystems. The best review of the theory and its consequences is provided by MacArthur (1972).

Several decades of laboratory studies of competition among species in a uniform environment providing a single resource have demonstrated that the elimination of all but one species is the usual result. Therefore, the persistence of many species in nature must result in large part from (a) the presence of a range of resources and (b) environmental heterogeneity. Theoretical considerations suggest that the number of competing species cannot exceed the number of distinct resource types or the number of distinct combinations of them. If the environment is absolutely unvarying, there is no theoretical limit to the similarity (i.e., overlap) of competing species but fluctuations in resource availability make very similar species vulnerable to competitive exclusion. Therefore, in fluctuating environments, coexisting competing species must differ in their ecological requirements. Empirically, Diamond (1973) has found that in many coexisting New Guinea birds that differ primarily in size, the ratio between the weights of the larger bird and the smaller bird averages 1.90 and is never less than 1.33 or more than 2.73. Species differing in size by a lesser amount are found to segregate by habitat and/or interspecific territoriality. There is also less complete evidence from other taxonomic groups that competitive interactions are important in determining community patterns [for example tropical bats (McNab, 1971a) and African ungulates (Lamprey, 1963, 1964)].

B. Predation, Parasitism, and Disease

The results of predator-prey interactions are different depending upon whether the predator has a threshold of prey abundance below which it cannot harvest prey often enough to meet its basic metabolic requirements, or whether the predator is capable of completely exterminating the prey. In the latter case, oscillations in predator-prey abundances tend to be amplified and the predator-prey system becomes one in which local extinctions and emigrations are responsible for the general overall patterns of distribution and abundance (Huffaker, 1970). In the former case, the prey are vulnerable to predation only when above threshold densities and are not exterminated even locally by their predators. The importance of spatial distributions for predation rates on tropical tree seedlings has been demonstrated by Connell (1971).

In general, we can say that the oscillations inherent in predator-prey interactions are damped when (a) the predator has a high threshold below which it cannot exploit the prey, (b) the predator has traits, such as territoriality, which prevent build-up of dense populations irrespective of prey densities, (c) there is hyperparasitism or hyperpredation such that the survival of offspring of the predator is strongly density-dependent, and (d) the environment is heterogeneous, thereby making it more difficult, in terms of time and energy, for the predators to find and exploit the prey.

Disease and parasites, as a special case, influence populations by causing direct mortality, reducing natality, and interfering with attainment of optimal physical condition. A successful parasite must invade the host and lodge in a cell, tissue, or organ where the conditions and substrates for development and/or reproduction are available. The parasite or its propagules must leave the host and survive until a new opportunity for host invasion occurs. An individual may die of his infection, become chronically infected, or, in many diseases, live to become immune. In the case of death or immunity that individual is no longer a suitable environment for the parasite. As more and more hosts in the population die or are rendered immune, the absolute number of susceptible hosts declines and the epidemic subsides. If the probability of the parasite being transmitted from one host to the next falls to too low a level, the disease will disappear from the population. Alternatively, the disease may continue to be transmitted within the population at constant low (endemic) levels and/or may produce periodic new epidemics as new susceptible individuals are recruited into the population through birth, movement, or loss of prior immunity (Paul, 1958).

Parasites and infectious disease agents that have been interacting over a long period of time evolve a relationship in which the host species is not seriously compromised, thus assuring the agent of harborage in the future. High host-mortality rates are ecologically unstable in instances where the disease-causing agent is dependent upon a given host system for maintenance.

Parasites have evolved a wide variety of strategies for entering their hosts, multiplying, and then affecting transmission. Since the most troublesome pathogens of animals and man in the tropics occur in modified ecosystems, little study has been devoted to parasitism in undisturbed situations. We might speculate that species diversity and low host population density in primary ecosystems would select pathogens which were host specific, as *Leishmannia* (Lainson and Show, 1970) and *Salmonella* (Lins, 1970) and/or for more efficient transmission systems which utilize host or vector behavior for seeking out new hosts, such as biting arthropods, or ingestion of prey by predators. Studies of a number of viruses (Horsfall and Tamm, 1965), protozoans, and helminths (Soulsby, 1965; Davis et al., 1970; Davis and Anderson, 1971) suggest that many tropical parasites have these life-history strategies. We are not certain that parasite species diversity is greater in the tropics than in temperate zones but of 107 known arthropod-borne viruses of animals, 75 are tropical (Subcommittee on Information Exchange of the American Commission on Arthropod-borne Viruses, 1971).

The fact that populations growing in a limited environment regularly show an S-shaped growth curve indicates that competition effects are apparent at population densities far below the maximum that can be sustained by the available resources. In fact, the growth rates of populations in limited environments deviate from the maximum possible exponential growth rates at any number greater than zero. This means that the activities of a predator depressing the population below the maximum do not eliminate competition. Rather, they gradually decrease its intensity. Close competitors, however, are highly vulnerable to extinction by selective predators and predation and therefore may act to reinforce the tendency of competition to force species to become more different.

For those species that compete directly for space, different dynamics result because space, unlike food, is not progressively reduced in quality by exploitation. Instead, up to virtual exhaustion of space, growth can be almost exponential. Consequently, a given intensity of predation reduces the intensity of competition much more drastically. It is significant that the cases in which increasing predation results in an increase in prey species richness are those in which the prey are actively competing for space, i.e., sessile intertidal animals (Paine, 1966) and green plants (Gillett, 1962; Janzen, 1970a). Also, for herbivores, the size of food particles (leaves) is much less significant than the physical and chemical characteristics of the tissues that determine the rate at which energy can be harvested from the plants. Finally, since grazing rates can strongly influence competitive interactions among plants (Harper, 1969), there may be powerful, although as yet poorly understood, feedback loops among predation, competition, and community structure. Theoretical and empirical analysis will have to emphasize these kinds of interactions in the future.

C. Mutualism

Mutualistic interactions have usually evolved from predator or parasite and host interactions (Smith, 1968). One member of the interaction increases the stability of resource levels for the second. The benefits deriving from this energy commitment include activities of the second, which increase the fitness of the first. Among these activities are pollen transfer, seed dispersal, and defense of vegetative tissues. Most terrestrial examples of mutualistic interaction involve plants and animals because plants are sessile and have greater need for the services provided by mobile animals.

The importance of mutalistic interactions apparently increases from temperate to tropical climates. For example, there are no obligate ant-plant mutualisms north of 24°, no nectarivorous or frugivorous bats north of 32 to 33°, and no orchid bees north of 24°. Extrafloral nectar glands on plants drop off drastically between the northern limits of the neotropics in northeastern Mexico and Texas (Gilbert, unpublished data). Also within the tropics, mutualistic interactions are more prevalent in the warm, wet evergreen forests than in the cooler and more seasonal habitats. The reasons for such trends seem to involve the fact that as one proceeds toward uniformly moist tropical conditions the year-round impact of insect herbivores on plants increases substantially. This, in turn, puts stronger selection pressure on plants to evolve various means of escape from predators, such as spines, hairs, and secondary chemicals. Coevolutionary responses to herbivores (see Ehrlich and Raven, 1965 and Gilbert, 1971 for discussions) theoretically are faster for outbreeding rather than inbreeding plant species. If potential pollinators exist, animal pollination should regularly be favored over wind pollination because the stronger specificity that is possible between animals and plant increases the efficiency of pollen transfer in terms of the proportion of grains produced that arrive at a conspecific stigma.

Another force that would select for a higher degree of specificity is the fact that offspring of interspecific crosses contain chemical cues used by the host-specific predators of both parent species, providing a genetic bridge for the predators of one species to invade the other.

Tropical pollination biology has received considerable attention and has been well reviewed (Faegri and van der Pijl, 1971). In general, we know that wind-pollinated species are a small proportion of the total and are largely restricted to disturbed habitats. At high elevations insect pollinators become less important and birds more important (Cruden, 1972; Heinrich and Raven, 1972) presumably because insects operate inefficiently at the low temperatures and are therefore less reliable as pollinators. Long-distance pollination by euglossine bees (Janzen, 1971b) and hummingbirds (Wolf, Hainsworth, and Stiles, 1972) is associated with plants that produce one to a few flowers at a time but over a long season. This strategy provides enough energy to support defense of the plant. Highly specific pollination systems are abundant, particularly in the lowland tropics (Dodson et al., 1969; Gilbert, 1972), and many tropical species obligately outcross in spite of their hyperdispersion as individuals.

A different area of mutualistic interaction involves seed dispersal. Whereas efficient pollinators allow plants to escape evolutionarily from predators, seed-dispersal agents allow spatial and temporal escape from seed and seedling predators that feed on concentrations of these juvenile stages of plants. The interactions of plants, their seed (or seedling) predators, and their dispersal agents are thought to be important in generating the high diversity of plant species in the tropics (Janzen, 1970a).

Plant-herbivore theory is currently the subject of active research. The general nature of plant defenses against herbivores is known, and in a few cases the selective activity of herbivores has actually been measured (Cates, 1971; Jones, 1962, 1966; Feeny, 1970). For the tropics we know that understory shrubs tend to be high in chemical defenses as they are the source of most of the world's spices (Baker,

1964). Early successional species, like their temperate-zone counterparts, appear to have generally fewer defenses than later successional and climax species (Cates and Orians, unpublished manuscript). The growth of these species in agricultural monocultures is therefore a system especially vulnerable to decimation by herbivores (Janzen, 1970b; Pimentel, 1961; Tahvanainen and Root, 1972) and hence this system and its dynamics are among the most fundamental in terms of their applied significance. Presently our information on the following basic aspects of tropical plant-herbivore interactions is extremely inadequate: (a) kinds of defensive chemicals produced by tropical plants; (b) relationship of chemical defenses to habitat type, successional status, and potential yield of useful products to man; (c) degree of generalization or specialization in the diets of tropical herbivores, especially insects; (d) dietary and habitat predilections of herbivores that are the most effective invaders of early successional habitats and agricultural crops; and (e) the influence of plant defenses on higher trophic levels in the community (Brower, 1969; Brower and Brower, 1964). All of these questions are worthy of research.

VI. Community Consequences of Population Interactions

A. Patterns of Species Diversity

It is known that the number of animal and plant species is for most groups greater in tropical areas than in areas at other latitudes. Many biologists believe that the tropics are also the place of origin of most plant and animal groups (Darlington, 1957; Moynihan, 1971), although the concept needs re-examining. Many data indicate that the ecological and behavioral relationships among tropical organisms are much more complex than among organisms in other areas. Much of the problem in understanding the reasons for this rich diversity lies in separating cause from effect. For example, are there more species in the tropics because of the subtle and complex adjustments in their ecology and behavior, e.g., reduced niche sizes, or are these complex interactions a *result* of the great diversity? Any approach to these problems must recognize this difficulty. Several theories have been proposed to explain the latitudinal gradient of increased species diversity as one approaches the equator, and these have been summarized by Baker (1970) and Pianka (1966). These theories include the following factors which may singularly, or in various combinations, explain the species richness of the tropics: (a) increased productivity, (b) greater spatial heterogeneity, (c) more intense competition, (d) higher rates of predation, and (e) lower extinction rates coupled to higher speciation rates. With the possible exception of the higher productivity, conclusive data on the other points are still lacking.

For example, we still do not know if the relative number of parasites and/or predators is greater in tropical as compared to nontropical locales.

The general problem of what regulates the range of a certain species may be mentioned here for it offers much promise for both theoretical and practical applications. It is possible that limits are imposed by biological interactions and a fundamental problem in community structure is to understand how competitive and predator-prey interactions differ between temperate and tropical areas. One promising line of approach might be to study the genetic, behavioral, and ecological characteristics of pairs or groups of species at the center of their respective ranges, and at their coterminal boundaries (see Wallace, 1959; Paine, 1966; MacArthur and Levins, 1967; MacArthur, 1969, 1972).

The community structure of birds would seem to be best understood, and appears to be explicable in terms of competitive interactions among the species. Some of the surprising gaps in the ranges between closely related pairs of species (Wilson, 1958; MacArthur, 1972; Diamond, 1972; Terbourgh, 1971) may be explained in this manner. The vertical compression of foraging heights characteristic of tropical forests as compared to temperate forest (MacArthur, Recher, and Cody, 1966; Orians, 1969a; Karr, 1971) can perhaps be explained in terms of contraction of foraging sites but not diet, as is predicted from optimal foraging theory in the face of efficient competitors (MacArthur and Pianka, 1966). Better data are required for all groups, but the apparent mesh of theory with data is encouraging. Another equally and not necessarily exclusive approach is the comparative one in which instead of analyzing one community in depth, several communities are compared in widely separated geographic areas. A particularly fruitful approach might well involve communities in both the New and Old World tropics.

Still another promising approach to these problems is exemplified by the theory of island biogeography, developed by MacArthur and Wilson (1967), tested primarily with tropical islands (Diamond, 1972; Karr, 1971) but with wide applicability to a variety of situations. This theory suggests that most islands are in a state of dynamic equilibrium between immigration and extinction of species and that diversity is determined by the island's present physical characteristics (size, distance from the mainland, and topographic relief) and independent for the most part from the island's history. This latter point may not always be true, and studies specifically directed to elucidating the possible role of "historic accident" ought to be encouraged, whether the population is insular or not. Islands constantly receive immigrants due to random dispersal, and this rate of immigration should increase with island size and with the proximity to the colonizing source. Island populations, like those in noninsular but contained areas, risk extinction because of competition and random fluctuations in numbers. Thus small islands with the smallest populations should have the highest extinction rates. When immigration and extinction rates are at equilibrium, large islands should have more species than those more remote. A logical consequence of this is that areas with many species should have relatively more rare species than those contained areas with fewer species. All of those predictions are remarkably in agreement with the available data.

These results also are important for mainland habitats that have "island" characteristics. Such habitats are becoming more and more frequent with the "checkerboard" destruction of the tropical forests by man. Páramo island community patterns have been investigated by Vuilleumier (1970), islands of pine forests in Central America by an Organization for Tropical Studies course in 1972, and individual plants and plant species as islands have received at least preliminary theoretical attention. These mainland islands differ from true islands in that the intervening areas are not as different as the ocean and hence species find it easier to cross the barriers. Related to this is the fact that organisms adapted to adjacent communities would find it easier to occupy the so-called island in the absence of its usual species than would be the case of an oceanic island. It would appear that local extinction rates are very high for tropical species since many of them exist at very low population densities. An increasing probability of extinction with decreasing population size is a basic feature of the successful models in the analysis of equilibrium systems. As yet we know nothing about this aspect of tropical population biology. More knowledge of it might enable us to design agricultural and other systems that would take advantage of these features rather than setting out to create conditions that would maximize local population densities, minimize extinction rates, and increase the probability of higher levels of infestation in the future. Also, the high extinction rates on small habitat islands should be taken into consideration during planning the sizes of and connections between tropical forest preserves.

Another striking characteristic of many tropical populations is the phenomenon of "patchiness" (MacArthur, 1972). Whereas the local presence or absence of temperate-zone species can generally be predicted from knowledge of their habitat requirements, many tropical species may be absent from a considerable fraction of the localities offering suitable habitat for them. The phenomenon appears to be real and its explanation is still lacking. One theory (Diamond, 1973) suggests that patchy distributions are due to lower dispersal rates of tropical species, which prolong the existence of distributional gaps as temporary phenomena, and the greater pressure from interspecific competition, which stabilizes gaps as indefinitely maintained phenomena.

B. Phenological Structure of Communities

Current theory suggests that in the temperate zones the physical environment generally appears to be the prime determinant of the time of year in which animals breed and plants flower. It further treats these extratropical environments, whether they are relatively stable or unstable in physical factors, as being more predictable. That is to say, the environmental clues signaling the onset of the physiologically permissive period for reproduction, for instance, are usually more abrupt and reliable than those available to tropical organisms. In tropical areas, the vicissitudes of the physical environment such as drought and heavy rains can and do exert a strong influence on the biota. Yet the available data, although meager, suggest that in the long run and in most years, biotic factors tend to exert the dominant influence in tropical areas. Janzen (1967) suggest that a great number of tree species flower during the dry season in Central America, not because it is the most permissive period physiologically in which to flower but because by using reserves at this time the tree least jeopardizes its competitive position. Some of these trees set fruit at a time when the fruits will have the highest probability of dispersal and/or the greatest freedom from predators.

The concepts of stability, harshness of the environment, and predictability, whereas very useful in theoretical terms, are poorly supported by data. Physical data are available for both the temperate and tropical areas, yet the strength of the correlation of the animal and plant activity patterns through time with these data are, for the most part, not known. Phenological studies of plant and animal activity patterns, especially those of long-term duration, would seem to be a research approach well worth encouraging in this regard. Such studies ought to clearly discriminate, in a philosophical sense, between proximate and ultimate factors. For example, the ultimate evolutionary reason why a tree flowers or fruits at a particular time is because, on the average, the maximum number of offspring result at that time as compared to flowering at some other time. The proximate reason or reasons may be perhaps simple physiological response to dryness or some other physical clue present in the period of the tropical dry season. Studies of both proximate and ultimate factors should be given equal weight; the former is of particular economic interest.

C. Patterns of Energy Flow

We have indicated that energy budgets represent the summation of the acquisition and expenditure patterns of individuals. Certain aspects of these patterns can be predicted from the physical environment and from the nature of the interactions of an individual with other individuals of the same and other species. The patterns of energy flow through populations and trophic levels in communities must be products of these individual energy budgets, and we wish to explore the extent to which analysis of individual energy budgets can provide insights about community-level patterns. Although we have a general knowledge of community primary production and know that it is strongly correlated with evapotranspiration (Rosenzweig, 1968), this statistic alone does not provide much insight into the relative allocations of energy resources to different activities such as growth, reproduction, and inter- and intraspecific interactions. We approach this problem via the concept of trophic levels, since organisms at different levels face sufficiently different problems that they are not conveniently lumped into single groups for the purposes of making generalizations.

Energy captured by photosynthesis can be shunted to such processes as vegetative growth, reproduction, or defense against enemies. Because sunlight is a directional and nonstorable resource, competitors can be deprived of it if a plant can grow taller and shade them. Therefore, it is not surprising that in climates with sufficient rainfall, selection favors plants that allocate energy to the production of woody supporting tissues that enable photosynthetic tissues to be located higher above the ground. Other things being equal, selection ought to favor the minimum diversion of energy to support consistent with withstanding the physical stresses of the environment and attacks of enemies, because more energy is thereby available for making energy gathering structures and reproductive materials. Therefore, from energy budget considerations, we would expect trees to allocate less energy to wood in regions with less wind and under circumstances with less attack by herbivores. As indicated by plant-herbivore theory and data, pressures are probably less on early successional species that may regularly escape in space and time than on climax species that are more predictable. Thus we can predict a greater range of wood types in tropical forests because early successional species in the wet tropics are subjected to minimal physical stresses and can grow tall with a small amount of wood. On the other hand, insect pressure in mature tropical forests appears to be higher than in temperate forests, suggesting that the commitment to defenses in tropical woods should be high. This prediction is supported by the existence of such early successional species in the wet tropics as *Cercropia* and *Ochroma*, which produce very light wood and have extremely rapid vertical growth, and the many primary forest species with extremely heavy, insect-resistant wood (Wolcott, 1957; Budowski, 1961). Nevertheless, the status of knowledge of investment of woody tissues in tropical trees characteristic of different successional stages is still very meager and many more data will be needed to establish these trends more clearly. It is known that tropical forests generally have higher ratios of leafy to woody biomass than temperate forests (Jordan, 1971) but there are no direct studies of the action of natural selection on these attributes.

In environments with little rainfall, herbaceous plants are commonly the dominant species, and these communities have very different patterns of energy flow. There is no long-term storage of energy in the form of cellulose and a higher percentage of the net annual primary production is consumed directly by herbivores. In the African tropics, these are chiefly large grazing mammals, whereas in the neotropics insects appear to be the major grazers, which is in part the result of a Pliocene-Pleistocene extinction of grassland ungulates in South America.

There is a second important influence of the physical environment on energy allocations by plants. At higher latitudes, where there is a much higher average wind velocity, many plants are wind pollinated and have wind-dispersed seeds. Such plants do not allocate any energy to mutualistic interactions with animals since animal vectors are not involved at any stage in the life history. Therefore the quantities and ranges of resources available for animals at all trophic levels are much less in communities dominated by wind-dispersed plants. In contrast, in tropical forests where wind pollinated species are rare, many species of animals are supported by pollen, nectar, and fruits, and community richness and the amounts of energy flowing to different trophic levels are strongly affected.

Animals allocate energy to growth, reproduction, predator avoidance, social interactions, and other processes. Some of these allocations express themselves in obvious physical structures. Less obvious, but perhaps energetically more important, is the commitment to a larger and more complex nervous system and sense organs. These information-processing systems are necessary if the organism is to react rapidly to predators, engage in complex social interactions, and find suitable prey hidden in a complex mosaic of species. Since natural selection favors predators that are more efficient in capturing prey, it selects for sensory investments on the part of the predators and the responses of the prey may be similarly affected. The more types of predators a prey is exposed to the greater the allocation from its energy budgets it may make. These commitments are often subtle, but as humans we are intuitively aware of the large commitment we make to our central nervous system, which consumes much more oxygen than would be expected strictly on the basis of its weight. It is possible that much of the energy lost between trophic levels can be accounted for in terms of coevolutionary interactions between organisms of different trophic levels.

We contend that (a) individuals are the units of selection and (b) evolution is a property of populations (i.e., changes in gene frequencies); hence causal explanations of community level patterns of energy flow must ultimately be derived from the energy budgets of individuals and the interactions at the population levels. Although higher level phenomena may be derived from the interactions we have described, it is clear that understanding of community-level problems will not result from studies that are simply oriented toward populations. Nevertheless, these studies need to be strongly grounded in a knowledge and understanding of the underlying mechanisms.

VII. Pest Population Management

A. Management Strategies in Agroecosystems

Agroecosystems were invented by man to satisfy the food and fiber needs of his expanding population centers. Forests, grasslands, and desert transformed to agroecosystems comprise a significant percentage of the earth's tillable surface. The process continues, collectively representing man's most disruptive effect on the world ecosystem. The myriad of pests that now threaten the stability of these agroecosystems are the products of these systems.

Both plant and animal pests challenged the progenitors of domesticated species long before the invention of agriculture, but counterselection pressures constrained their populations within the long-term carrying capacities of their environments and regulated the virulence of pathogens at moderate levels that would preserve the hosts. These dynamics have been upset by agricultural practices. Potential hosts are renewed periodically by man; consequently highly virulent pest species strains are selected. These colonize rapidly and reproduce abundantly, resulting in pest populations that often achieve highly destructive levels.

Agroecosystems have evolved over time to highly specialized, carefully regulated production systems. They often comprise vast areas and are composed of homogeneous, and often near homozygous, high-density populations that are highly selected for yield or energy-conversion efficiency. Theory and considerable experimental data (Tahvanainen and Root, 1972; Southwood and Way, 1970) suggest that such highly simplified ecosystems are more vulnerable to devastation by pests than the more genetically and species-diverse natural systems. In crop plants many epiphytotics have been attributed to the lack of diversity in these monocultures, the most recent being the corn blight outbreak in the USA in 1970 (Tatum, 1971). The hazards of this trend are likely to affect several important crop agroecosystems according to the recent report by the United States National Academy of Sciences (National Academy of Sciences, 1972). Nevertheless, the monoculture production system is unlikely to be abandoned because of the increasing demand for food and fiber. Some compromises must be discovered whereby ecosystem diversity can be enhanced without major reductions in productivity.

Since the host plants or animals are a resource renewed continually by man in his agroecosystems, the most effective natural mechanisms for limiting population levels and destructiveness are not available.

In response, man has developed many methods for restricting pest outbreaks. Important among these has been the use of chemicals. The undesirable side effects of some of these chemicals are well known, especially the inadvertant elimination of additional population regulators in the form of predators and parasites (Newsom, 1972). In some situations this approach has failed completely to achieve economical control of insect pests. This and many other factors have dictated a management approach to pest problems which integrates methods for one type of pest (e.g., insects) but that also integrates strategies for all pests affecting a given agroecosystem. The approach has been labeled "integrated pest management."

Effective pest management systems are essential to agriculture, especially in tropical areas. But before such practical systems can be developed, basic information must be obtained on population composition, population regulation (predators, parasites, and pathogens), and climatic effects on populations. Some of the necessary data are location-specific but many will have general applicability. These must be obtained under a systems approach to the study of tropical biology.

The pressure of change has never been greater in the neotropics because of the significant international and national programs aimed at "modernizing" so-called "traditional" tropical agriculture. Since the neotropics comprises most of the developing nations of the Western Hemisphere, neotropical agroecosystems reflect principally the characteristics of traditional agriculture.

Traditional agriculture is a production system evolving when the marginal productivity of labor is very low and the incentives to save are weak because the marginal productivity of capital is also very low (Schultz, 1964). It is labor intensive (and thus operates at low energy levels), low in productivity, and approaches a state of biological equilibrium. Further, the tillage units are generally small and sparsely planted with seeds of mixed genetic types (Sanchez, 1972). Such a "mixed cultural system" is not as readily exploitable by endemic plant pests as modern monoculture systems. It also provides some protection against climatic adversity and attack by new pests because of its inherent heterogeneity and genotypic recombination potential. The tillage system of traditional agriculture is generally so poor that the plants are not as susceptible to some pests as types grown under more favorable conditions. It is probable that the traditional agricultural system has greater influence on the regulation of plant pest populations at the present time than the unique features of the neotropical environment.

The last decade has witnessed an unparalleled revolution in the development, dissemination, and adoption of new agricultural technology (Dalrymple, 1972; Borlaug, 1971). The rapid extension of this technology to developing nations is now referred to as the "green revolution." Whether one agrees or disagrees with the rubric "green revolution" to refer to this phenomenon (Paddock, 1970), it is a revolution sparked by experiences with plantation agriculture fueled by the application of genetic principles to the development of high-yield, pest-resistant crop varieties (resistant to some pests but in some instances more susceptible to others). The agroecosystem changes that are coincident with this revolution pose highly significant implications for pest management (Wilkes and Wilkes, 1972; Apple, 1972; Smith, 1972).

Modern man has developed the capability for a highly efficient agricultural production system. Genetic materials have been manipulated to achieve high-yielding, homogeneous populations of either homozygous or heterozygous crop plants. But in this achievement, it has been necessary to abort many of the biological mechanisms or habitat characteristics basic to the survival of the progenitors of cultivated crop plants. For example, (a) seed dormancy characteristics have been eliminated, (b) the products of inter- and intra-population breeding have been avoided by "certified" or "pure seed" production procedures, (c) dispersion among and competition with other plant species have been eliminated with herbicides and/or tillage, (d) propagule dissemination mechanisms have been eliminated, (e) genetic diversity has been greatly reduced on a worldwide basis by controlled breeding programs (National Academy of Sciences, 1972), and (f) many inherent biochemical and physical mechanisms endowing pest resistance have been lost by chance selection. Many of these events and other similar ones have rendered modern monoculture production systems more vulnerable to devastating pest attack than traditional systems. Other attributes such as irrigation, multiple cropping, high rates of fertilization, and tillage (Yarwood, 1968) have enhanced vulnerability to some pests. Highly controlled agroecosystems based on monocultures are creations of the twentieth century and are recent innovations in the history of agriculture. Episodes of pest devastation within highly managed agroecosystesm based on monocultures dictate caution and suggest more comprehensive surveillance systems to forewarm of pest outbreaks.

Although ecosystem diversity does not insure ecosystem stability, logic and much observational evidence support the thesis that diversity does enhance stability (Southwood and Way, 1970). Modern agroecosystems are less diverse than traditional ones in many ways: genetically homogeneous crops, clean culture by tillage and/or use of herbicides, and possible control of selected insects or diseases with chemicals (some beneficial, some harmful). As Harlan (1961) and others have pointed out, the genetic diversity of cultivated plants in traditional agricultural systems probably is due to the introgressive hybridization of cultigens with companion, closely related weed species. Modern agriculture is rapidly eliminating these introgressive, dynamic breeding systems in the developing world, a source of potential crop improvement genes for the future. As Harlan (1961) recognized, "A predicament now exists in which the technologically backward countries cannot afford to keep their great varietal resources and the more progressive countries cannot afford to let them be discarded. The only answer is an extensive exploration and collection program devoted to assembling as much of the germ plasm of the world as possible and the diligent maintenance of the material once it is obtained."

Another apparent fact of the tropics is the greater number of pest species and the propensity for diseases in the tropics as compared to the temperate zone. Wellman (1969) has made observations that support this thesis.

The neotropics exemplifies both modern and traditional agricultural forms. Modern agriculture was introduced into the tropics in the form of plantation culture in the latter nineteenth and early twentieth centuries. Three of these major crops—banana, cacao, and rubber—are propagated clonally; consequently, large areas were planted to genetically homogeneous populations. Numerous highly destructive diseases have threatened these crops continuously in the neotropics (Wellman, 1972). Such problems as panama disease *(Fusarium oxysporum, F. cubense)* and moko disease *(Pseudomonas solanacearum)* of bananas, witches broom *(Marasmius perniciosus)* and pod rots *(Monilia* sp. and *Phytophthora palmivora)* of cacao, and South American leaf blight *(Dothidella ulei)* of rubber have cost the plantation operators in the neotropics and the consumers of their production hundreds of millions of dollars.

These diseases have also dictated the production areas for these crops and have brought great financial hardship to areas forced out of production. These experiences in the neotropics have been biologically devastating and financially catastrophic, but they occurred on plantations under good management and with crop protection expertise available. An extension of the green revolution in the neotropics to the ultimate could create ecological conditions similar to but much more expansive than plantation agriculture.

The crop protection programs of the neotropical countries are wholely inadequate to sustain the green revolution (Apple and Smith, 1972; Echandi et al., 1973; Caltagirone et al., 1973). Even though traditional agriculture remains the dominant form, crop losses due to pests have already reached levels of intolerance in most countries. It is difficult to obtain accurate data on crop losses caused by pests anywhere in the world, but it is especially difficult in developing nations. Those who know the field situations understand that losses are exorbitant even though some of the recorded percentages may be exaggerated. Recently compiled statistics for Peru illustrate the high loss of potential crop production attributable to pests (Apple and Smith, 1972). The value of crop production in Peru for 1970 was estimated at $487 million. The Sub-Dirección de Defensa Fitosanitaria of the Peruvian Ministry of Agriculture estimated that 38.2 percent of the potential production in 1970 was lost due to pests with a monetary value of $187 million (assuming stable market prices). These losses by type of pest are illustrated in Table 2.1.

Although loss rates as high as these are alarming in food-deficient countries, the lack of a crop-protection response capability within ministries of agriculture and universities portends a dimmer future. The lack of this capability is partly due to lack of knowledge of the threat of pests to agricultural systems, lack of opportunities for trained people in crop-protection programs, lack of adequate quarantine programs or facilities, and inadequate relationships between research and application in the field.

It is essential that we develop an integrated approach to the containment of these pests. We must integrate the methods of population (or disease incidence) suppression into a pest management system employing actions consistent with a quality environment. This integration of methodology must occur not only within a crop protection discipline (e.g., entomology) but between disciplines to achieve a systems approach to the management of insect, disease, nematode, and weed pests of a given agroecosystem. For too many years, these disciplines have pursued control strategies independently. This independence cannot continue if we are to effect management systems based on ecological principles of population regulation discussed earlier in this section. Even our present rudimentary knowledge of competition, predation, and mutualism shows that manipulations of one group designed to manage pests such as soil-borne nematodes could interfere with the control strategies for insects or other organisms. Thus, there must be an integration of methods within and between disciplines to achieve the desired objectives. We have not done this effectively anywhere in the world. But there may be better opportunities for initiating these types of management systems in the neotropics where disciplinary parochialism (as manifested through scientists' guilds) is less well entrenched than in the U.S. or other developed countries. Finally, it is clear that successful pest manage-

TABLE 2.1. Crop losses by type of pest

Type of pest	Crop loss (%)	Crop loss (U.S.$-millions)
Insects	15.8	77.3
Diseases	10.8	53.0
Nematodes	6.0	29.3
Weeds	5.6	27.4
Totals	38.2	187.0

ment must be based on a firm ecological knowledge of populations.

B. Management Strategies for Forest Pests

Concepts and methods of pest management in agroecosystems discussed in the introduction to this section are largely pertinent to managed natural and man-made (plantation) forest ecosystems. Major differences can be attributed principally to the prolonged rotation and the greatly increased height dimension (vegetational structure) of managed forest ecosystems. These characteristics, however, may affect the choice of pest management alternatives.

Because forest management in tropical America is in its infancy (Kalkkinen, 1960) phytophagous insects as pests of forests have not received the intensive study that they have as pests of agriculture (Gray, 1972). The purpose of this subsection is to emphasize the need for initiating basic ecological research on the potential insect and disease pests most likely to jeopardize the success of present and planned forest-management practices.

Forest ecosystems in the neotropics are under ever increasing pressure as sources for water, fuel, and wood products (Wadsworth, 1971). Clearly, reforestation of cleared forest land and intensive management of existing forests to increase productivity provide the most practical means of satisfying both local and regional demands for forest resources. As sustained yield forestry practices are adopted, a sharp increase in the frequency and severity of insect and disease problems should be anticipated. In managed tropical forests, the vegetational and animal richness is often drastically reduced. By favoring a few desirable tree species and minimizing stand age diversity, wood production is increased and inherent problems of harvesting and utilization common to natural mixed stands are overcome (DeBoer, 1969; FAO Staff, 1960). These same conditions, however, favor the increase of destructive insects (Graham, 1969; Pimentel, 1961). Insect outbreaks in mixed tropical forests have been witnessed by Kalshoven (1953) in Indonesia, by Brereton (1957) in Australia, and by Gray (1972) in New Guinea. Characteristically these outbreaks occurred in forests with impoverished flora or in situations in which a single tree species was common. The severe depredations of *Dendroctonus* bark beetle on native pine forests of Central America (Beal, 1965; Schwerdtfeger, 1955) provide further evidence of the instability of simple forest ecosystems. Gray (1972) emphasizes that insect outbreaks on single host species or individuals in mixed tropical forests are fairly common. Although of little impact in natural forests, such localized perturbations may have significant economic importance in managed forests due to the high commercial value of indigenous tree species.

Plantations of fast-growing tree species may be the most feasible means of reforesting the extensive acreages of land that have been taken out of production by shifting agriculture (FAO Staff, 1960).

Indeed, there may remain little choice because regeneration of humid tropical forests once destroyed may be extremely slow or may never occur. Yet, planted forests, particularly monocultures of exotic tree species, have proven notoriously susceptible to devastation by native herbivores in many eastern tropical countries (Beeson, 1941; Brown, 1962, 1965; Curry, 1965; Dhanarajan, 1969; Gray, 1968; Roberts, 1969). Furthermore, there is no reason to believe that neotropical plantations will be endowed with unique immunity. On the contrary, several insect pests already have caused concern in experimental plantations. The mahogany shoot borer, *Hypsipyla grandella*, is a particularly noxious pest of planted Meliaceae throughout the American tropics (Martorell, 1943; Ramirez Sanchez, 1964; Sliwa, 1968). A variety of native insects, particularly defoliating caterpillars, has caused recent concern in plantations of introduced pine and cypress (Bustillo, 1970; Bustillo and Lara, 1972; Vélez, 1966) in Colombia. Outbreaks of scale insects have been observed on exotic trees in Puerto Rico (DeLeon, 1941; Martorell, 1939-1940; Wolcott, 1940) and various species of phytophagous insects have been recorded on exotic plantations in Costa Rica (Gara, 1970). Due to the lack of periodic surveys, doubtless many other herbivore outbreaks have gone undocumented.

In a review of pan-tropical forest entomology, Gray (1972) summarized the status of our knowledge of tropical forest insect pests. A list of forest-tree diseases in Latin America has been compiled by Gárces (1964). Unlike in many eastern tropical countries, permanent forest-pest research programs in Latin America are either nonexistent or are of very recent origin. Thus, there is a paucity of information on the biology and host relationships of all but a few herbivorous pests. Only the mahogany shoot borer, termites, ambrosia beetles, and pine bark beetles have received more than cursory study by visiting forest entomologists.

The mahogany shoot borer constitutes the major obstacle encountered to date in the establishment of plantations of Meliaceae. Larvae of this insect bore into the stems and terminal shoots of young plants, particularly those belonging to the genera *Cedrela* (spanish cedar) and *Swietenia* (mahogany). As a result, seedlings either are killed or severely distorted (Ramirez Sanchez, 1964). In recent years, a considerable volume of literature has appeared on *H. grandella* principally as a result of intensive research programs in progress at IICA in Turrialba, Costa Rica.

Eventual protection of Meliaceae plantations from *H. grandella* may result from one of the following approaches: (a) biological control (Bennett, 1969; Rao and Bennett, 1968, (b) silvicultural control (Sliwa, 1968; Ramirez Sanchez, 1964), (c) genetic resistance of host plants (Grijpma and Ramalho, 1969; Grijpma, 1970), and (d) sustained release of systemic chemicals (Allen, Gara and Wilkens, 1970; Wilkens, 1972). The last two alternatives are especially promising.

Termites are well-known pests in the tropics and probably destroy more seasoned timber than any other group of insects (Gray, 1972; Snyder and Zetek, 1924; Wolcott, 1946, 1954). Certain species of termites are also capable of causing considerable damage to commercial tree species in natural and planted forests in Central and South America (Browne, 1968; FAO Staff, 1960; Harris, 1966a). Economic losses to standing timber are almost wholely due to termites that possess internal protozoa that digest cellulose—the dry-wood termites (Kalotermitidae) and the damp-wood termites (Rhinotermitidae and Termopsidae) (Harris, 1966a). Damage to live mahogany and pine by termites of the genus *Coptotermes* has been reported in British Honduras and Guatemala (Harris, 1959, 1966a; Williams, 1965 a,b). As economic pests in neotropical regions, termites have been well studied and excellent papers and reviews by Harris (1965a,b) and bibliographies by Snyder (1956, 1961, 1968) are available. The ecological role of termites in tropical-forest ecosystems, however, remains poorly understood (Gray, 1972).

Ambrosia beetles (Coleoptera: Scolytidae and Platypodidae) are perennial and widespread pests of freshly felled logs, weakened or dying trees, and unseasoned timbers of the tropics. Their main damage results from the blackstained tunnels they construct in their host material. Browne (1961) and Kalshoven (1958, 1959) conducted extensive studies of eastern tropical Scolytidae. Doubtless much of this information on biology and host selection behavior is pertinent to western tropical species. Ventocilla's (1965) studies of environmental factors influencing flight activity of *Xyleborus ferrugineus* and the work of Samaniego and Gara (1970) on flight habits and host selection behavior of *Xyleborus* spp. and *Platypus* spp. constitute the limited behavioral knowledge of neotropical Ambrosia beetles. Devastating outbreaks of *Dendroctonus* bark beetles (Coleoptera: Scolytidae) have occurred in native pine forests of Central America (Beal, 1965; Schwerdtfeger, 1955); and *Ips* bark beetles, also pine forest pests, have been studied briefly by Billings (1973).

There is also a need for fundamental research on the many other species of herbivore pests now being encountered in the Latin American tropics as the biology and potential economic impacts of most species is unknown. Indeed, the lack of basic research on population ecology of forest insect pests has greatly hindered their intelligent and effective management, both in temperate (Belyea, 1965) and tropical (Gray, 1972) forests.

In tropical forest ecosystems, the ultimate goals of research should be to elucidate mechanisms by which herbivore populations are regulated in natural mixed and single genera forests in order to discover the causes of insect outbreaks. This knowledge, in turn, will provide an ecological basis for the development of silvicultural practices that minimize economical losses from pests in managed forest ecosystems.

Before these goals can be realized, however, the following more basic needs are prerequisites:

1. Comprehensive surveys of managed forests and experimental forest plantations throughout the neotropics to identify and evaluate present and potential pest problems.
2. Studies on the life histories, biologies, and host relationships of herbivorous species that are expected to present obstacles to tropical forest regeneration and management efforts.
3. Evaluations of the applicability of temperate pest management methods (Beroza and Knipling, 1972; Graham, 1959; Knipling, 1966; Vite, 1970; Voûte, 1964) to tropical forest pest problems.

C. Management Strategies for Animal and Human Diseases

Infectious and parasitic diseases impose constraints on populations of both wild and domesticated animals and on man himself. Although diseases are not unique to the tropics, some of the most devastating diseases of man and his animals do occur there, and are a consequence of the development of agroecosystems. As noted above, diseases that seriously affect the host population occur in changing ecosystems, and are the result of introductions of pathogens or hosts, or situations leading to increased population densities of host or vector species. There are several well-documented examples of effects of the establishment of new host-pathogen associations as the result of modification of ecosystems for agricultural use.

One of the most dramatic examples of a new host-pathogen association resulting from introduction of the pathogen, with devastating consequences for the host, is rinderpest in Africa. This virus disease swept Africa for the first time in the 1890s affecting populations of cattle and large wild mammals, and has remained there (Scott, 1970). The impact of rinderpest has been great. Present-day distributions of several native species has been determined by past epizootics (Pearsall, 1954), with persisting effects on animal populations and economic life (Spinage, 1962; Pankhurst, 1966). Foot-and-mouth disease (Rossiter and Albertyne, 1949; Anon., 1963) and brucellosis (Witter and O'Meara, 1970; Rollinson, 1962; Heisch et al., 1963) pose a problem of a different kind in African wildlife. *Brucella* spp. affect populations of ungulates through abortion and by sterility in both sexes, both of which can significantly reduce natality. Perhaps the greatest potential danger to the large African herbivores is the possibility of an attempt to implement disease control in domesticated livestock by slaughter of the wildlife reservoir species, practiced in some areas for control of trypanosomiasis (Wells and Lumsden, 1970).

Disease problems in tropical areas have resulted from the introduction of new host or vector species, either casually or more permanently, into existing host-pathogen systems. The worker who enters the forest to cut trees, gather chicle, or hunt may contract sylvan yellow fever (Soper et al., 1934) or leishmaniasis (Lainson and Shaw, 1970). Okinawan settlers attempting to colonize lowland forests of Bolivia had to abandon their well-organized enterprise when decimated by "jungle fever" (Schaeffer et al., 1959; Schmidt et al., 1959). Some areas of Africa are unsuitable for human occupation or cattle raising because of trypanosomiasis (sleeping sickness and nagana), a benign protozoan disease in large, wild ungulate reservoirs (Wells and Lumsden, 1970). African swine fever, a virus of wild porcines producing a highly fatal disease in domesticated swine, renders some areas unsuitable for pig production (De Kock et al., 1940; De Tray, 1960; Henschele and Coggins, 1965). In established livestock-raising areas in the American tropics, rabies in cattle transmitted by vampire bats (Constantine, 1970) is among the most serious of livestock diseases (World Health Organization, 1966), resulting in an estimated annual loss of 100 million dollars (Steele, 1966), and up to 80 percent loss in some areas (Ruiz Martinez, 1963). In addition to the economic loss in cattle, rabies also represents a human health hazard. Although vampire-transmitted rabies in the Americas probably dates to pre-Columbian times, the introduction and subsequent increase of cattle greatly expanded the availability of vampire food leading to a gradual increase in vampire populations and an intensification of livestock rabies over time (Johnson, 1971).

Rapidly expanding human populations result in increasing colonization of lowland tropical areas, and doubtless will continue over the foreseeable future. More than a million people are establishing new settlements in the lowland tropics in Latin America alone (The Institute of Ecology, 1972). In his perturbation of the original ecosystems, man has often inadvertantly created ecological conditions favoring transmission of diseases to himself or his animals, either by creating conditions under which vectors or intermediate hosts increase or by increasing host population density to the point where transmission is facilitated.

Several troublesome tropical diseases are insect-borne. Agricultural activities have simplified tropical forest ecosystems, reduced diversity, and greatly expanded the larval habitats of certain vector insects. Forest clearing in the mountains and hills of Malaya resulted in greatly increased populations of an important malaria vector, *Anopheles maculatus*. Coconut and rubber plantations in the same area favor other malaria vectors, *A. letifer* and *A. barbirostris*. Creation of lowland ricefields and ponds makes extensive larval habitat for *A. aconitis, A. philippensis,* and the *A. hyrcanus* group (Sandosham, 1959). Paddy-rice culture throughout much of Asia resulted in seasonally large populations of Vishnui *Culex* mosquitoes; the two most abundant, *C. tritaeniorhynchus* and *C. gelidis,* are very effieient vectors of Japanese encephalitis virus. When large populations of these mosquitoes are juxtaposed with an exzootic wildlife maintenance cycle of the virus, and populations of susceptible swine, the amplifying host, epidemics occur with transmission to man and losses of pigs (Heathcote, 1970; Bendell, 1970; Simpson et al., 1970; Hill, 1970).

There is a clear trend toward increasing flock/herd size in modern agriculture. High host-population densities in the presence of vector insects may result in an explosive epizootic if a pathogen is introduced. Over the past 40 years, periodic epizootics of Venezuelan Equine Encephalitis (VEE) occurred in northern South America. The last of these was the most severe, lasting two years (1969 to 1971), and swept an area from Ecuador to Texas (Groot, 1972). Mortality rates in horses, mules, and burros were high, and substantial numbers of fatal human cases occurred in these areas (Avilan, 1964). The variety of mosquito vector species that transmit the virus has always been present. Increased equine populations set the stage for the epidemic.

Perhaps one of the best examples of the trend toward large populations of homogeneous vertebrates is the modern poultry industry, where single flocks of more than a million individuals are common. Contact transmission of pathogens among uniformly susceptible vertebrate hosts of high population density has resulted in devastating epizootics. The world poultry industry is now recuperating from such a panzootic—viscerotropic Asiatic Newcastle disease (Hanson et al., 1973).

The potential of disease associated with high population densities is not confined to domestic species. It is fair to say that all species, both wild and domesticated, have natural pathogens, and are susceptible to infection by certain pathogens of other species. Shortage of animal protein for human consumption prompted interest in commercial production and utilization of game species. Both large African antelopes (Crawford, 1967; Bindernagel, 1968; Swank, 1972) and the South American capybara (Ojasti, 1971; Ojasti and Medina, 1972), the world's largest rodent, are being considered for commercial game cropping. Managers will have to be alert to the possibility of emerging disease problems as population densities of these species increase.

The establishment of new host-pathogen relationships will have to be avoided to control infectious disease of man and animals. Where disease problems exist, all factors that impinge upon the host-parasite relationship must be defined to determine which ones are most vulnerable to attack. Field epidemiological research is justified by the prospect of disease control through accrual of information for devising a control strategy, given the proposition that control is more involved than dispensing pills, brewing up batches of vaccines, and spraying a few thousand more hectares with pesticides.

The practical value of a broad biological approach to control is well illustrated by the studies on vampire bats, which led to successful trials through the use of anticoagulants (Thompson et al., 1972), and control of the screw worm fly by the release of sterile males (Knipling, 1960). Many other studies are needed. Population dynamics, ethology, and interactions of wild African ungulates, coupled with additional studies of immunity, pathogenesis, and transmission will have to be carried out before serious thoughts of control of rinderpest or foot-and-mouth disease in nature can be entertained. Similarly, wildlife-domestic animal virus relationships for foot-and-mouth disease in South America is unstudied, despite the millions of dollars that are going into control of this disease in livestock in most of the infected countries.

The diversity of lowland tropics, compounded by the role of arthropods, makes an understanding of the epidemiology of insect-borne disease doubly difficult. For example, environmental tolerances and preferences of arthropod vectors, as they influence rates of pathogen multiplication in the vector and distribution, longevity, and behavior of the vector are known to be important (Cavanaugh and Marshall, 1972) but are generally poorly studied even in important diseases such as Venezuelan and Japanese encephalitis and the dengue fevers of man.

The principles of integrated pest management (Rabb, 1970, 1972) apply to reduction of disease in animal and human populations with one important exception. Crop pest managers strive for maintenance of pests at low levels. Reduction to lower levels in diseases potentially threatening the existence of an entire animal industry, as rinderpest, or threatening human health, as viral encephalitis, may not be an acceptable strategy, however, and attempts at control or even eradication may be required.

References

Allan, G. G., R. I. Gara, and R. M. Wilkins. 1970. Studies on the shootborer *Hypsipyla grandella* Zeller. III. The evaluation of some systemic insecticides for the control of larvae in *Cedrela odorata L.* Turrialba 20: 478-487.

Anon. 1963. Foot and mouth disease in non-domesticated animals. Bull. Epizoot. Dis. Africa II: 143-146.

Antonovics, J. 1966. Evolution in adjacent populations. Unpublished Ph.D. Thesis, University of North Wales.

Antonovics, J. 1972. Population dynamics of the grass *Anthoxanthum odoratum* on a zinc mine. J. Ecol. 60: 351-365.

Apple, J. L. 1972. Intensified pest management needs of developing nations. BioScience 22: 461-464.

Apple, J. L., and R. F. Smith. 1972. A preliminary study of crop protection problems in selected Latin American countries. Report for Agency for Int. Develop. Contract No. AID/csd 3296. 41 p.

Avilan, J. 1964. El brote de encefalitis equina Venezolana al norte del estado Zulia a fines de 1962. Rev. Venez. Sanidad Asst. Soc. 29: 231-236.

Ayala, F. J., J. R. Powell, M. L. Tracey, C. A. Mourão, and S. Pérez-Salas. 1970. Enzyme variability in *Drosophila willistoni* group. IV. Genic variation in natural populations of *Drosophila willistoni*. Genetics 70: 113-139.

Baker, H. G. 1964. Plants and civilization. Wadsworth, Belmont, California. 194 p.

Baker, H. G. 1970. Evolution in the tropics. Biotropica 2: 101-111.

Beal, J. A. 1965. Bark beetles threaten destruction of Honduras pine forests. FAO/IUFRO Symp. on Internationally Dangerous Forest Diseases and Insects. Oxford, 1964. Vol. 1, parts 11/111. FAO, Rome. 3 p.

Beeson, C. F. C. 1941. The ecology and control of the forest insects of India and the neighboring countries. Vasant Press, Dehra Dun, 1st reprint. 1961. 767 p.

Belyea, R. M. 1965. The role of the forest entomologist. FAO/IUFRO Symp. on Internationally Dangerous Forest Diseases and Insects. Oxford, 1964. Vol. 1. FAO, Rome. 5 p.

Bell, C. R. 1970. Seed distribution and germination experiment, p. D-177-D-182. In H. T. Odum and R. F. Pigeon (eds.) A tropical rain forest. A study of irradiation and ecology at El Verde, Puerto Rico. U.S. Atomic Energy Commission, Oak Ridge, Tennessee.

Bennett, F. D. 1969. Report on a cursory survey of the natural enemies of *Hypsipyla grandella* in British Honduras, June-July, 1968. Commonwealth Inst. Biol. Control, Trinidad. 7 p.

Bendell, P. J. E. 1970. Japanese encephalitis in Sarawak: Studies on mosquito behavior in a land Dyak village. Trans. Roy. Soc. Trop. Med. Hyg. 64: 497-502.

Beroza, M., and E. F. Knipling. 1972. Gypsy moth control with the sex attractant pheromone. Science 177: 19-27.

Billings, R. F. 1973. Colonization patterns of two species of *Ips* (Coleoptera: Scolytidae) co-habiting pines in British Honduras and Honduras, 11 p. *In* O. T. S. 72-2 Ecology of Central American Pine Forests. Organization for Tropical Studies, San José, Costa Rica.

Bindernagel, J. A. 1968. Game cropping in Uganda. A report on an experimental project to utilize populations of wild animals for meat production in Uganda, East Africa. Uganda Game Dept., Entebbe. Mimeo. 200 p.

Boer, P. J. den, and G. R. Gradwell (eds.) 1971. The dynamics of populations. Proc. Adv. Study Inst. on Dynamics of Numbers in Populations. Oosterbeek, Netherlands, Wageninger, Pudoc. 611 p.

Borlaug, N. E. 1971. Mankind and civilization at another crossroad. McDougall Memorial Lecture. FAO Conference, Rome, Nov. 1971. FAO, Rome.

Brereton, L. J. 1957. Defoliation in rain-forest. Australian J. Sci. 19: 204-205.

Brower, L. P. 1969. Ecological chemistry. Sci. Amer. 220(2): 22-29.

Brower, L. P., and J. V. Z. Brower. 1964. Birds, butterflies, and plant poisons: A study in ecological chemistry. Zoologica 49: 137-159.

Brown, J. L. 1964. The evolution of diversity in avian territorial systems. Wilson Bull. 76: 160-169.

Brown, J. L., and G. H. Orians. 1970. Spacing patterns in mobile animals. Ann. Rev. Ecol. Syst. 1: 239-262.

Brown, K. W. 1962. Notes on the recent outbreaks of looper caterpillars in coniferous plantations in Uganda. Uganda Forest. Dept. Tech. Note No. 99. p. 1-7.

Brown, K. W. 1965. Insect damage on exotic conifers. Proc. 12th. Int. Congr. Entomol. London, 1964. Academic Press, New York. 677 p.

Browne, F. G. 1961. The biology of Malayan Scolytidae and Platypodidae. Gov. Print. Office, Malay Forest Rec. No. 22. Kuala Lumpur. 255 p.

Browne, F. G. 1968. Pests and diseases of forest plantation trees. Oxford Univ. Press, London. 1330 p.

Budowski, G. 1961. Studies on forest succession in Costa Rica and Panama. Unpublished Ph.D. Thesis, Dept. Forestry, Yale University, New Haven, Conn. 189 p.

Bustillo, A. E. 1970. Gusano defoliador del ciprés. Bol. Divulgación No. 31. Inst. Colombiano Agropecuário, Medellin, Colombia. 12 p.

Bustillo, A. E., and L. Lara. 1972. Plagas forestales. Bol. Divulgación No 33. Inst. Colombiano Agropecuario, Medellin, Colombia. 32 p.

Caltagirone, L. E., J. R. Orsenigo, W. J. Kaiser, and M. Allen. 1973. The crop protection situation in Guatemala, Honduras, Nicaragua, Costa Rica, Panama and Guyana. Report for Agency for Int. Development. Contract No. AID/csd 3296. Washington, D.C. 74 p.

Canfield, R. H. 1957. Reproduction and life span of some perennial grasses of southern Arizona. J. Range Manage. 10: 199-203.

Carter, G. S., and L. C. Beadle. 1931a. The fauna of the swamps of the Paraguayan Chaco in relation to its environment. I. Physico-chemical nature of the environment. J. Limn. Soc. London 37: 205-258.

Carter, G. S., and L. C. Beadle. 1931b. The fauna of the swamps of the Paraguayan Chaco in relation to its environment. II. Respiratory adaptations in fishes. J. Limn. Soc. London 37: 327-366.

Cates, R. G. 1971. The interface between slugs and wild ginger: Some evolutionary aspects. Unpublished Ph.D. Thesis. Dept. of Botany, University of Washington, Seattle. 94 p.

Cavanangh, D. C., and J. D. Marshall. 1972. The influence of climate on seasonal prevalence of plague in the Republic of Vietnam. J. Wildl. Dis. 8: 85-94.

Clark, L. R., P. W. Geier, R. D. Hughes, and R. F. Morris. 1967. The ecology of insect populations in theory and practice. Methuen, London. 232 p.

Cody, M. L. 1966. A general theory of clutch size. Evolution 20: 174-184.

Cole, L. C. 1954. Population consequences of life history phenomena. Quart. Rev. Biol. 29: 103-137.

Cohen, D. 1966. Optimizing reproduction in a randomly varying environment. J. Theor. Biol. 12: 119-129.

Cohen, D. 1967. Optimizing reproduction in a randomly varying environment when a correlation may exist between the conditions at the time the choice has to be made and the subsequent outcome. J. Theor. Biol. 16: 1-14.

Cohen, D. 1968. A general model of optimal reproduction in a randomly varying environment. J. Ecol. 56: 219-228.

Connell, J. H. 1970. A predator-prey system in the marine intertidal region. I. *Balanus glandula* and several predatory species of *Thais*. Ecol. Monogr. 40: 49-78.

Connell, J. H. 1971. On the role of natural enemies in preventing competitive exclusion in some marine animals and in rain forest trees. *In* P. J. den Boer and G. R. Gradwell (eds.) The dynamics of populations. Proc. Adv. Study Inst. on Dynamics of Numbers in Populations. Oosterbeek, Netherlands.

Constantine, D. G. 1970. Bat rabies: Current knowledge and future research, p. 253 to 275. *In* Y. Nagano, and F. M. Davenport (eds.) Rabies. Proc. Working Conf. on Rabies. University Park Press, Baltimore, Maryland.

Corbet, P. 1962. A biology of dragonflies. Witherby, London. 247 p.

Crawford, M. A. 1967. Possible use of wild animals as future sources of food in Africa, p. 83-87. *In* Proc. 85th. Ann. Congr. Brit. Vet. Assoc. Southport, England.

Crook, J. H. 1965. The adaptive significance of avian social organization. Symp. Zool. Soc. London 14: 181-218.

Cruden, R. W. 1972. Pollinators in high-elevation ecosystems: Relative effectiveness of birds and bees. Science 176: 1439-1441.

Curry, S. J. 1965. The incidence and control of indigeneous insect pests of Kenya forest plantations, with particular reference to wood borers in exotic crops. Proc. 12th. Int. Congr. Entomol. London, 1964. p. 678-679. Academic Press, New York.

Da Cunha, A. B., H. Burla, and T. Dobzhansky. 1950. Adaptive chromosomal polymorphism in *Drosophila willistoni*. Evolution 4: 212-235.

Dalrymple, D. G. 1972. Imports and plantings of high-yielding varieties of wheat and rice in the less developed countries. FED Report No. 14, Foreign Econ. Develop. Serv., U.S. Dept. Agric. Washington, D.C. 56 p.

Darlington, P. 1957. Zoogeography. The geographical distribution of animals. Wiley, New York. 675 p.

Davis, J. W., and R. C. Anderson (eds.). 1971. Parasitic diseases of wildlife mammals. Iowa State University Press, Ames. 364 p.

Davis, J. W., L. H. Karstad, and D. O. Trainer (eds.). 1970. Infectious diseases of wild mammals. Iowa State University Press, Ames. 421 p.

DeBoer, D. 1969. Impressions of industrial problems in tropical forestry, p. 79-84. In O.T.S. Introduction to Tropical Forestry. Organization for Tropical Studies, San José, Costa Rica.

DeKock, G., E. M. Robinson, and J. J. G. Keppel. 1940. Swine fever in South Africa. Onderstepoort. J. Vet. Sci. Anim. Ind. 14: 31-93.

DeLeon, D. 1941. Some observations on forest entomology in Puerto Rico. Caribbean Forest 2: 160-163.

DeTray, D. E. 1960. African swine fever - an interim report. Bull. Epizoot. Dis. Afr. 8: 217-223.

Dhanarajan, G. 1969. Future of Malayan forest entomology. Malayan Forest. 32: 61-66.

Diamond, J. M. 1972. Comparison of faunal equilibrium turnover rates on a tropical island and a temperate island. Proc. Nat. Acad. Sci. USA 68: 2742-2745.

Diamond, J. M. 1973. Distributional ecology of New Guinea birds. Science 179: 759-769.

Dodson, C. H., R. L. Dressler, H. G. Hills, R. M. Adams, and N. H. Williams. 1969. Biologically active compounds in orchid fragrances. Science 164: 1243-1249.

Downhower, J. F., and K. B. Armitage. 1971. The yellow-bellied marmot and the evolution of polygamy. Amer. Natur. 105: 355-370.

Echandi, E., M. Shenk, G. T. Weekman, J. Knoke, and E. L. Nigh. 1973. Overview of crop protection problems in Brazil, Uruguay, Bolivia, Ecuador, and the Dominican Republic. Report, Agency Inter. Develop. Contract No. AID/csd 3296, in preparation.

Ehrlich, P. R., and L. E. Gilbert. 1973. The population structure and dynamics of a tropical butterlfy, *Heliconius ethilla.* Biotropica 5: 69-82.

Ehrlich, P. R., and R. W. Holm. 1962. Patterns and populations. Science 137: 652-657.

Ehrlich, P. R., and P. H. Raven. 1965. Butterflies and plants: A study in coevolution. Evolution 18: 586-608.

Emlen, J. M. 1966. The role of time and energy in food preferences. Amer. Natur. 100: 611-617.

FAO Staff. 1960. Practicas de plantación forestal en America Latina. FAO Cuadernos de Fomento Forestal No. 15. FAO, Rome. 500 p.

Faegri, K., and L. van der Pijl. 1971. The principles of pollination ecology. Pergamon, Oxford. 291 p.

Feeny, P. 1970. Seasonal changes in oak leaf tannins and nutrients as a cause of spring feeding by winter moth caterpillars. Ecology 51: 565-581.

Fleming, T. H. 1970. Comparative biology of two temperate-tropical rodent counterparts. Amer. Midl. Natur. 83: 462-471.

Fleming, T. H. 1971. Population ecology of three species of neotropical rodents. Misc. Publ. Mus. Zool. Univ. Mich. 143: 1-77.

Fournier, L. A. 1969. Observaciones preliminares sobre la variación altitudinal en el numero de famílias de arboles y de arbustos en la vertiente del Pacífico de Costa Rica. Turrialba 19: 548-552.

Gadgil, M., and W. H. Bossert. 1970. Life historical consequences of natural selection. Amer. Natur. 104: 1-24.

Gara, R. I. 1970. Report of forest entomology consultant. Interamer. Inst. Agr. Sci., OAS, Turrialba, Costa Rica. UNDP Project 80: 1-21

Garcés, O. C. 1964. Las enfermedades de los árboles forestales en la America Latina y su impacto en la producción forestal. Symp. FAO/UIOIF Internationally Important Insects and Diseases. July, 1964. Sect. IV. FAO, Rome. 9 p.

Gilbert, L. E. 1971. Butterfly-plant coevolution: Has *Passiflora adenopoda* won the selectional race with Heliconiine butterflies? Science 172: 585-586.

Gilbert, L. E. 1972. Pollen feeding and reproductive biology of *Heliconius* butterflies. Proc. Nat. Acad. Sci. USA 69: 1403-1407.

Gilbert, L. E., and M. C. Singer. 1973. Dispersal and gene flow in a butterfly species. Amer. Natur. 107: 58-72.

Gillett, J. B. 1962. Pest pressure, an underestimated factor in evolution, p. 37 to 46. *In* D. Nichols (ed.) Taxonomy and geography. Systematics Assoc. Publ. No. 4. London.

Graham, S. A. 1959. Control of insects through silvicultural practices. J. Forest. 57: 281-283.

Grant, P. R. 1968. Polyhedral territories of animals. Amer. Natur. 102: 75-80.

Gray, B. 1968. Forest tree and timber insect pests in the Territory of Papua and New Guinea. Pacific Insects 10: 301-323.

Gray, B. 1972. Economic tropical forest entomology. Ann. Rev. Entomol. 17: 313-354.

Grijpma, P. 1970. Immunity of *Toona ciliata* M. Roem. var. *australis* (F.V.M.) C.D.C. and *Khaya ivorensis* A. Chev. to attacks of *Hypsipyla grandella* Zeller in Turrialba, Costa Rica. Turrialba 20: 85-93.

Grijpma, P., and P. Ramalho. 1969. *Toona* spp. posibles alternatives para el problema del barrenador *Hypsipyla grandella* de las Meliaceae en America Latina. Turrialba 19: 531-547
Groot, H. 1972. The health and economic impact of Venezuelan equine encephalitis (VEE). Venezuelan Encephalitis. Pan. Am. Health Org. Publ. Sci. 243: 7-16.
Guevara, S., and A. Gómez-Pompa. 1972. Seeds from surface soils in a tropical region of Veracruz, Mexico. J. Arnold Arboretum 53: 312-335.
Hainsworth, F. R., and L. L. Wolf. 1972. Power for hovering flight in relation to body size in hummingbirds. Amer. Natur. 106: 589-596.
Haffer, J. 1969. Speciation in Amazonian forest birds. Science 165: 131-137.
Hammel, H. T. 1964. Terrestrial animals in cold: recent studies of primitive man, p. 413-434. *In* D. B. Dill (ed.) Handbook of Physiology. Amer. Physiol. Soc. Washington, D. C.
Hammen, T. van der, and E. Gonzalez. 1960a. Upper Pleistocene and Holocene climate and vegetation of the "Sabana de Bogota" (Colombia, S. A.) Leidse Geol. Mededelinger 25: 261-315.
Hammen, T. van der, and E. Gonzalez. 1960b. Holocene and late glacial climate and vegetation of Paramo de Palacio (Eastern Cordillera, Colombia, S. A.). Geol. Mijnbouw 39: 737-745.
Hanson, R. P., J. Spalatin, and G. S. Jacobson. 1973. The viscerotropic pathotype of Newcastle disease virus. Avian Dis. 17: 354-361.
Harlan, J. R. 1961. Genetic origin of plants useful to agriculture, p. 3 to 21. *In* R. E. Hodgson (ed.) Germ plasm resources. Publ. No. 66, Amer. Assoc. Adv. Sci., Washington, D. C. 381 p.
Harper, J. L. 1967. A Darwinian approach to plant ecology. J. Ecol. 55: 247-270.
Harper, J. L. 1969. The role of predation in vegetational diversity, p. 48-62. *In* Diversity and stability in ecological systems. Brookhaven Symp. Biol. No. 22. Brookhaven Nat. Lab., Upton, New York.
Harper, J. L., and J. N. Clatworthy. 1963. The comparative biology of closely related species. VI. Analysis of the growth of *Trifolium repens* and *T. fragiferum* in pure and mixed populations. J. Exp. Bot. 14: 172-190.
Harper, J. L., and J. H. McNaughton. 1962. The comparative biology of closely related species living in the same area. VII. Interference between individuals in pure and mixed populations of *Papaver* species. New Phytol. 61: 175-188.
Harris, W. V. 1959. Notes on termites injurious to forestry in British Honduras. Empire Forest. Rev. 38: 181-185.
Harris, W. V. 1966a. The role of termites in tropical forestry. Insectes Sociaux 13: 255-265.

Harris, W. V. 1966b. Termites and trees. A review of recent literature. Forest Abstr. 27: 173-178.
Hartshorn, G. S. 1972. Ecological life history and population dynamics of *Pentaclethra macroba* a tropical wet forest dominant and *Stryphuodendron excelsum,* an occasional associate. Unpublished Ph.D. Dissertation, University of Washington, Seattle.
Heathcote, O. H. U. 1970. Japanese encephalitis in Sarawak: Studies on juvenile mosquito populations. Trans. Roy. Soc. Trop. Med. Hyg. 64: 483-488.
Heinrich, B., and P. H. Raven. 1972. Energetics and pollination ecology. Science 176: 597-602.
Heisch, R. B., E. R. N. Cooke, A. E. Harvey, and F. De Souza. 1963. Isolation of *Brucella suis* from rodents in Kenya. E. Afr. Med. J. 40: 132-133.
Henschele, W. P., and L. Goggins. 1965. Isolation of African swine fever virus from a giant forest hog. Bull. Epizoot. Dis. Afr. 13: 255-256.
Hill, M. N. 1970. Japanese encephalitis in Sarawak: Studies on adult mosquito populations. Trans. Roy. Soc. Trop. Med. Hyg. 64: 489-496.
Holdridge, L. R. 1967. Life zone ecology. Tropical Science Center, San José, Costa Rica. 206 p.
Holdridge, L. R., W. C. Grenke, W. H. Hatheway, T. Liang, and J. A. Tosi Jr. 1971. Forest environments in tropical life zones: A pilot study. Pergamon Press, New York. 747 p.
Horn, H. 1971. The adaptive geometry of trees. Princeton University Press, Princeton, New Jersey. 155 p.
Horsfall, F. L. Jr., and I. Tamm (eds.) 1965. Viral and rickettsial infections of man. Lippincot, Philadelphia. 1282 p.
Howard, H. E. 1920. Territory in bird life. J. Murray, London. 308 p.
Huffaker, C. B. (ed.) 1970. Biological control. Proc. Amer. Assoc. Adv. Sci. Symp. Biol. Control. Plenum Press, New York. 511 p.
Hutchinson, G. E. 1959. Homage to Santa Rosalia, or why are there so many kinds of animals? Amer. Natur. 93: 145-159.
Inger, R. F. 1959. Temperature responses and ecological relations of two Bornean lizards. Ecology 40: 127-136.
Inger, R. F., and J. P. Bacon 1968. Annual reproduction and clutch size in rain forest frogs from Sarawak. Copeia 1968: 602-606
Inger, R. F., and B. Greenberg. 1966. Annual reproductive patterns of lizards from a Bornean rain forest. Ecology 47: 1007-1021.
Janzen, D. H. 1967. Synchronization of sexual reproduction of trees within the dry season in Central America. Evolution 21: 620-637.
Janzen, D. H. 1969. Seed-eaters versus seed size, number, toxicity and dispersal. Evolution 23: 1-27.

Janzen, D. H. 1970a. Herbivores and the number of tree species in tropical forests. Amer. Natur. 104: 501-528.

Janzen, D. H. 1970b. The unexploited tropics. Bull. Ecol. Soc. Amer. 51: 4-7.

Janzen, D. H. 1971a. Seed predation by animals. Ann. Rev. Ecol. System. 2: 465-492.

Janzen, D. H. 1971b. Euglosine bees as long distance pollinators of tropical plants. Science 171: 203-205.

Janzen, D. H. 1972. Escape in space by *Stercalia apetala* seeds from the bug *Dysdercus fasciatus* in a Costa Rican diciduous forest. Ecology 53: 350-361.

Janzen, D. H. 1973a. Sweep samples of tropical foliage insects: Description of study sites, with data on species abundance and size distributions. Ecology 54: 659-686.

Janzen, D. H. 1973b. Sweep samples of tropical foliage insects: Effects of seasons, vegetation types, elevation, time of day and insularity. Ecology 54: 687-708.

Johnson, H. N. 1971. General epizootiology of rabies, p. 237 to 251. *In* Y. Nagano and F. M. Davenport (eds.) Rabies. Proc. Working Conf. on Rabies. University Park Press, Baltimore, Maryland.

Jones, D. A. 1962. Selective eatings of the acyanogenic form of the plant *Lotus corniculatus* L. by various animals. Nature 193: 1109-1110.

Jones, D. A. 1966. On the polymorphism of cyanogenesis in *Lotus corniculatus*. I. Selection by animals. Can. J. Genet. Cytol. 8: 556-567.

Jordan, C. F. 1971. A world pattern in plant energetics. Amer. Sci. 59: 425-433.

Kalkkinen, E. 1960. Forestry progress in Latin America, p. 229-232. In Fifth World Forestry Congress University of Washington, Seattle.

Kalshoven, L. G. E. 1953. Important outbreaks of insect pests in the forests of Indonesia. Trans. 9th. Int. Congr. Entomol. Amsterdam 1951: 229-234.

Kalshoven, L. G. E. 1958. Studies on the biology of Indonesian Scolytoidea. 4. Data on the habits of Scolytidae. 1st. part. Tijdschr. Entomol. 101: 157-180.

Kalshoven, L. G. E. 1959. Studies on the biology of Indonesian Scolytoidea. 4. Data on the habits of Scolytidae. 2nd. part. Tijdschr. Entomol. 102: 135-173.

Karr, J. R. 1971. Structures of avian communities in selected Panama and Illinois habitats. Ecol. Monogr. 41: 207-233.

Kennedy, J. S. 1956. Phase transformation in locust biology. Biol. Rev. 31: 349-370.

Knipling, E. F. 1960. Eradication of the screw-worm fly. Sci. Amer. 203(4): 54-61.

Knipling, E. F. 1966. New horizons and the outlook for pest control, p. 455-470. *In* Scientific Aspects of Pest Control. Nat. Acad. Sci., Nat. Res. Council Publ. No. 1402.

Lack, D. L. 1954. The natural regulation of animal numbers. Clarendon Press, Oxford. 343 p.

Lack, D. L. 1968. Ecological adaptations for breeding in birds. Methuen, London. 409 p.

Lainson, R., and J. J. Shaw. 1970. Leishmaniasis in Brazil. V. Studies on the epidemiology of cutaneous leishmaniasis in Mato Grosso State, and observations on two distinct strains of Leishmania isolated from man and forest animals. Trans. Roy. Soc. Trop. Med. Hyg. 64: 654-667.

Lamprey, H. F. 1963. Ecological separation of the large mammal species in the Tarangire Game Reserve, Tanganyika. E. Afr. Wildlife J. 1: 63-92.

Lamprey. H. F. 1964. Estimation of the large mammal densities, biomass and energy exchange in the Tarangire Game Reserve and the Masai Steppe in Tanganyika. E. Afr. Wildlife J. 2: 1-46.

Levins, R. 1968. Evolution in changing environments. Some theoretical explorations. Princeton University Press, Princeton, New Jersey. 120 p.

Levins, R. 1969. Thermal acclimation and heat resistance in *Drosophila* species. Amer. Natur. 103: 483-499.

Lins, C. C. 1970. Studies on enteric bacterias in the lower Amazon region. I. Stereotypes of *Salmonella* isolated from forest animals in Pará State, Brazil. Trans. Roy. Soc. Trop. Med. Hyg. 64: 439-443.

MacArthur, R. H. 1968. Selection for life tables in periodic environments. Amer. Natur. 102: 381-383.

MacArthur, R. H. 1969. Patterns of communities in the tropics. Biol. J. Linn. Soc. London 1: 19-30.

MacArthur, R. H. 1972. Geographical ecology. Pattern in the distribution of species. Harper and Row, New York. 269 p.

MacArthur, R. H., and R. Levins. 1967. The limiting similarity, convergence, and divergence of coexisting species. Amer. Natur. 101: 377-385.

MacArthur, R. H., and E. R. Pianka. 1966. On optimal use of a patchy environment. Amer. Natur. 100: 603-609.

MacArthur, R. H., H. Recher, and M. Cody. 1966. On the relation between habitat selection and species diversity. Amer. Natur. 100: 319-325.

MacArthur, R. H., and E. O. Wilson. 1967. The theory of island biogeography. Princeton University Press, Princeton, New Jersey. 203 p.

Marrero, J. 1943. A seed storage study of some tropical hardwoods. Caribbean Forest. 4: 99-106.

Martorell, L. F. 1939-40. Some notes on forest entomology. Caribbean Forest. 1(1): 25-26; 1(2): 31-32; 1(3): 23-24.
Martorell, L. F. 1943. Forests and forest entomology. Caribbean Forest. 4: 132-134.
McNab, B. K. 1963. Bioenergetics and the determination of home range size. Amer. Natur. 97: 133-140
McNab, B. K. 1969. The economics of temperature regulation in neotropical bats. Comp. Biochem. Physiol. 31: 227-268.
McNab, B. K. 1970. Body weight and the energetics of temperature regulation J. Exp. Biol. 53: 329-348.
McNab, B. K. 1971a. The structure of tropical bat faunas. Ecology 52: 352-358.
McNab, B. K. 1971b. On the ecological significance of Bergmann's rule. Ecology 52: 845-854.
Miller, A. H. 1963. Seasonal activity and ecology of the avifauna of an American equatorial cloud-forest. Univ. Calif. Publ. Zool. 66: 1-78.
Mooney, H. A., and M. West. 1964. Photosynthetic acclimations of plants of diverse origin. Amer. J. Bot. 51: 825-827.
Moreau, R. E. 1952. The place of Africa in the palearctic migration system. J. Anim. Ecol. 21: 250-271.
Moreau, R. E. 1963. Vicissitudes of the African biomes in the late Pleistocene. Proc. Zool. Soc. London 141: 395-421.
Moreau, R. E. 1966. The bird faunas of Africa and its islands. Academic Press, New York. 424 p.
Moynihan, M. 1971. Success and failure of tropical mammals and birds. Amer. Natur. 105: 371-383.
National Academy of Sciences, Agricultural Board. 1972. Genetic vulnerability of major crops. Washington, D. C. 307 p.
Newsom, L. D. 1972. Some ecological implications of two decades of use of synthetic organic insecticides for control of agricultural pests in Louisiana, p. 439 to 466. *In* M. T. Farvar and J. P. Milton (eds.) The careless technology. Natural History Press, Garden City, New York.
Ojasti, J. 1971. El chiguire. Defensa de la naturaleza 1(3): 4-14.
Ojasti, J., and G. Medina-P. 1972. The management of capybara in Venezuela. Trans. 37th. N. Amer. Wildl. Conf., p. 268-277. Wildlife Management Institute, Washington, D. C.
Orians, G. H. 1969a. The number of bird species in some Costa Rican forests. Ecology 50: 783-806.
Orians, G. H. 1969b. On the evolution of mating systems in birds and mammals. Amer. Natur. 103: 589-603.
Orians, G. H., and G. Collier. 1963. Competition and blackbird social systems. Evolution 17: 449-459.
Owen, D. F. 1971. Tropical butterflies: The ecology and behavior of butterflies in the tropics with special reference to African species. Claredon Press, Oxford. 214 p.

Paddock, W. C. 1970. How green is the green revolution? BioScience 20: 897-902.

Paine, R. T. 1966. Food web complexity and species diversity. Amer. Natur. 100: 65-75.

Paul, J. R. 1958. Clinical epidemiology. University of Chicago Press, Chicago, Ill. 291 p.

Pankhurst, R. 1966. The great Ethiopian famine of 1888-1892; a new assessment. J. Hist. Med. 21: 95-124.

Pearsall, W. H. 1954. Biology and land use in East Africa. New Biol. 17: 9-26.

Pianka, E. R. 1966. Latitudinal gradients in species diversity: A review of concepts. Amer. Natur. 100: 33-46.

Pimentel, D. 1961. Species diversity and insect population outbreaks. Ann. Entomol. Soc. Amer. 54: 76-86.

Pittendrigh, C. S. 1950. The ecoclimatic divergence of *Anopheles bellator* and *A. homunculus*. Evolution 4: 43-64.

Rabb, R. L. 1970. Introduction to the conference, p. 1-5. *In* R. L. Rabb and F. E. Guthrie (eds.) Concepts of pest management. Proc. of a Conference. North Carolina University Raleigh, N. C., March 25-27, 1970.

Rabb, R. L. 1972. Principles and concepts of pest management, p. 6-29. *In* Implementing practical pest management strategies. Proc. Nat. Insect-Pest Manage. Ext. Workshop. Purdue University, Lafayette, Ind., March 14-16, 1972.

Rabotnov, T. A. 1956. The life cycle of the *Heracleum sibiricum* L. Byull. Moskovskoe Obshchestov Ispytaletei Prirody, Odtel Biolog. 61: 73-80.

Rabotnov, T. A. 1958. The life cycle of *Ranunculus acer* L. and *R. auricomus* L. Byull. Moskovskoe Obshchestov Ispytaletei Prirody, Odtel Biol. 63: 77-86.

Ramirez Sanchez, J. 1964. Investigación preliminar sobre biólogia, ecólogia, y control de *Hypsipyla grandella* Zeller. Bol. Inst. Forest. Latino Amer. Invest. Capacitacion 16: 54-77.

Rao, V. P., and F. D. Bennett. 1968. Possibilities of biological control of the Meliaceous shoot borers *Hypsipyla* spp. 9th. Brit. Commonw. Forest. Conf., New Delhi, 1968. Commonw. Forest. Inst., Oxford, p. 1-14.

Raven, P. H., B. Berlin, and D. E. Breedlove. 1971. The origins of taxonomy. Science 174: 1210-1213.

Ricklefs, R. 1969. An analysis of nesting mortality in birds. Smithsonian Contrib. Zool. No. 9. 45 p.

Roberts, H. 1969. Forest insects of Nigeria with notes on their biology and distribution. Commonw. Forest. Inst., Dept. Forestry, Oxford, No. 44, p. 1-206.

Rollinson, D. H. L. 1962. *Brucella* agglutinins in East African game animals. Vet. Rec. 74: 904.

Rosenzweig, M. L. 1968. Net primary productivity of terrestrial communities: Prediction from climatological data. Amer. Natur. 102: 67-74.

Rossiter, L. W., and A. A. L. Albertyn. 1949. Foot and mouth disease in game. J. S. Afr. Vet. Med. Assoc. 18: 16-19.

Ruibal, R. 1961. Thermal relations of five species of tropical lizards. Evolution 15: 98-111.

Ruiz Martinez. C. 1963. Epizootiologia y profilaxis regional de la rabia paralítica en las Americas. Monografias Ediciones Protinal, Caracas.

Sagar, G. R. 1959. The biology of some sympatric species of grassland. Unpublished Ph.D. Thesis, University of Oxford.

Sagar, G. R. 1970. Factors controlling the size of plant populations. Proc. 10th. Brit. Weed Control Conf. 3: 965-979.

Salisbury, E. J. 1942. The reproductive capacity of plants: Studies in quantitative biology. G. Bell and Sons, London. 244 p.

Sanchez, P. 1972. Upland rice improvement under shifting cultivation systems in the Amazon Basin of Peru. Tech. Bull. No. 210. North Carolina Agric. Expt. Sta., North Carolina State Univ., Raleigh. 20 p.

Sandosham, A. A. 1959. Malariology with special reference to Malaya. University of Malaya Press, Singapore. 327 p.

Sarukhán, J. 1968. Analisis sinecologico de las selvas de *Terminalia amazonia* en la planície costera del Golfo de Mexico. Colegio de Postgraduados Escuela Nacional de Agricultura, Chapingo, Mexico.

Sarukhán, J. 1971. Studies on plant demography. Unpublished Ph.D. Thesis, Univ. of Wales.

Sarukhán, J. 1972. Demographic studies on grassland weed species. *In* Solomon and Price-Jones (eds.) Increasing the biological contribution to the control of pests and diseases. Blackwells, Oxford.

Sarukhán, J. 1974. Studies on plant demography *Ranunculus repens* L., *R. bulbosus* L. and *R. acris* L. II. Reproductive strategies and seed population dynamics. J. Ecol., in press.

Sarukhán, J. and M. Gadgil. 1974. Studies on plant demography *Ranunculus repens* L., *R. bulbosus* L., and *R. acris* L. III. Mathematical models and comparative demography. J. Ecol., in press.

Sarukhán, J., and J. L. Harper. 1973. Studies on plant demography *Ranunculus repens* L., *R. bulbosus* L., and *R. acris* L. I. Population flux and survivorship. J. Ecol. 61: 675-716.

Schaeffer, M., D. C. Gajdusek, A. Brown-L, and H. Eichenwald. 1959. Epidemic jungle fevers among Okinawan colonists in the Bolivian rain forests. I. Epidemiology. Amer. J. Trop. Med. Hyg. 8: 372-396.

Schmidt, J. R., D. C. Gajdusek, M. Schaeffer, and R. H. Gorrie. 1959. Epidemic jungle fever among Okinawan colonists in the Bolivian rain forests. II. Isolation and characterization of Uruma virus, a newly recognized human pathogen. Amer. J. Trop. Med. Hyg. 8: 479-487.

Schoener, T. W. 1967. The ecological significance of sexual dimorphism in size in the lizard *Anolis consperus*. Science 155: 474-477.

Schoener, T. W. 1968. The Anolis lizards of Bimini: Resource partitioning in a complex fauna. Ecology 49: 704-726.

Schoener, T. W. 1971. Theory of feeding strategies. Ann. Rev. Ecol. Syst. 2: 369-404.

Schoener, T. W., and D. H. Janzen. 1968. Notes on environmental determinants of tropical versus temperate insect size patterns. Amer. Natur. 102: 207-224.

Scholander, P. F., W. Flagg, V. Walters, and L. Irving. 1953. Climatic adaptations in arctic and tropical poikilotherms. Physiol. Zool. 26: 67-92.

Schultz, T. W. 1964. Transforming traditional agriculture. Yale University Press, New Haven, Conn. 212 p.

Schwerdtfeger, F. 1955. Informe al gobierno de Guatemala sobre la entomologia forestal de Guatemala. Vol. II. La plaga de *Dendroctonus* en los bosques de pinos y modo de combaterla. FAO/ETAP No. 366. FAO, Rome.

Scott, G. R. 1970. Rinderpest, p. 20-35. *In* J. W. Davis, L. H. Karstad and D. O. Trainer (eds.) Infectious diseases of wild mammals. Iowa State University Press, Ames.

Selander, R. K. 1965. On mating systems and sexual selection. Amer. Natur. 99: 129-141.

Simberloff, D. S., and E. O. Wilson. 1969. Experimental zoogeography of islands. The colonization of empty islands. Ecology 50: 278-296.

Simberloff, D. S., and E. O. Wilson. 1970. Experimental zoogeography of islands. A two-year record of colonization. Ecology 51: 934-937.

Simpson, D. I. H., E. T. W. Bowen, G. S. Platt, H. Way. C. E. G. Smith, S. Peto, S. Kamanth, L. B. Liat, and L. T. Wah. 1970. Japanese encephalitis in Sarawak: Virus isolation and serology in a land Dyak village. Trans. Roy. Soc. Trop. Med. Hyg. 64: 503-510.

Sliwa, D. D. 1968. *Hypsipyla* shoot borer activity in young mahogany and Spanish cedar plantations in Costa Rica. An exploratory study. 38 p. (cyclostyled). IICA-CIDIA, Turrialba, Costa Rica.

Slud, P. 1960. The birds of finca "La Selva," Costa Rica: A tropical wet forest locality. Bull. Amer. Mus. Natur. Hist. 121(2): 49-148.

Smith, C. C. 1970. The coevolution of pine squirrels *(Tamiasciurus)* and conifers. Ecol. Monogr. 40: 349-371.

Smith, N. G. 1968. On the advantage of being parasitized. Nature 219: 690-694.

Smith, R. F. 1972. The impact of the green revolution on plant protection in tropical and subtropical areas. Bull. Entomol. Soc. Amer. 18: 7-14.

Snow, D. W. 1962. A field study of the black-and-white manakin, *Manacus manacus*, in Trinidad. Zoologica 47: 65-104.

Snyder, T. E. 1956. Annotated, subject-heading bibliography of termites 1350 B. C. to A. D. 1954. Smithsonian Inst. Publ. No. 4258. 305 p.

Snyder, T. E. 1961. Supplement to the annotated, subject-heading bibliography of termites 1955-1960. Smithsonian Inst. Publ. No. 4463. 137 p.

Snyder, T. E. 1968. Second supplement to the annotated, subject heading bibliography of termites 1961-1965. Smithsonian Inst. Publ. No. 4705. 188 p.

Snyder, T. E., and J. Zetek. 1924. Damage by termites in the Canal Zone and Panama and how to prevent it. U.S. Dept. Agric. Bull. No. 1232: 1-26.

Soper, F. L., E. R. Rikard, and P. J. Crawford. 1934. The routine postmortem removal of liver tissue from rapidly fatal fever cases for the discovery of silent yellow fever foci. Amer. J. Hyg. 19: 549-566.

Soulsby, E. J. L. 1965. Textbook of veterinary clinical parasitology. F. A. Davis Co., Philadelphia, Pa. 1120 p.

Southwood, T. R. E., and M. J. Way. 1970. Ecological background to pest management, p. 6-29. *In* R. L. Rabb and F. E. Guthrie (eds.) Concepts of pest management. Proc. Conf. North Carolina State Univ., Raleigh, N. C., March 25-27, 1970.

Spinage, C. A. 1962. Rinderpest and faunal distribution patterns. Afr. Life 16: 55.

Steele, J. H. 1966. International aspects of veterinary medicine and its relation to health, nutrition and human welfare. Military Med. 131: 765-778.

Subcommittee on Information Exchange of the American Committee on Arthropod-Borne Viruses. 1971. Catalogue of arthropod-borne and selected vertebrate viruses of the world. Amer. J. Trop. Med. Hyg. 20: 1018-1050.

Swank, W. G. 1972. Wildlife management in Masailand, East Africa. Trans. 37th. N. Amer. Wildl. and Natur. Resources Conf., p. 278-286. Wildlife Management Institute, Washington, D.C.

Symington, C. F. 1933. The study of secondary growth on rain forest sites in Malaya. Malayan Forest. 2: 107-117.

Tahvanainen, J. O., and R. B. Root. 1972. The influence of vegetational diversity on the population ecology of a specialized herbivore, *Phyllotreta cruciferae* (Coleoptera: Crysomelidae). Oecologia 10: 321-346.

Tamm, C. O. 1948. Observations on reproduction and survival of some perennial herbs. Bot. Notiser 101: 305-321.
Tamm, C. O. 1956. Further observations on the survival and flowering of some perennial herbs. I. Oikos 7: 273-292.
Tatum, L. A. 1971. The southern corn leaf blight epidemic. Science 171: 1113-1116.
Terborgh, J. 1971. Distribution on environmental gradients: Theory and a preliminary interpretation of distributional patterns in the avifauna of the Cordillera Vilcabamba, Peru. Ecology 52: 23-40.
The Institute of Ecology. 1972. Man in the living environment. University of Wisconsin Press, Madison. 267 p.
Thompson, R. D., G. G. Mitchell, and R. J. Burns. 1972. Vampire bat control by systematic treatment of livestock with an anticoagulant. Science 177: 806-808.
Tinkle, D. W., H. M. Wilbur, and S. G. Tilley. 1970. Evolutionary strategies in lizard reproduction. Evolution 24: 55-74.
Tullock, G. 1971. The coal tit as a careful shopper. Amer. Natur. 105: 77-80.
Van der Pijl, L. 1969. Principles of dispersal in higher plants. Springer-Verlag, New York. 154 p.
Vázquez, C. 1973. Relación entre luz y germinación de semillas en especies secundárias de una selva húmeda, unpublished manuscript.
Vázquez-Yanes, C., and A. Gómez-Pompa. 1972. Ecofisiológia de la germinación de semillas de algunas especies del genero Piper. I Congr. Latinoamericano Mexicano Bot. Res. Trabajos (Abstr.).
Vélez, A. R. 1966. Nota sobre tres defoliadores del pino o ciprés (*Cupressus lusitanica* u *benthani* Mill.) en Antioquia. Agric. Trop. 22(12): 640-649.
Ventocilla, J. A. 1965. La influencia de la temperatura y la precipitación en la actividad de *Xyleborus ferrugineus* (F). Thesis Magister Scientiae. Inst. Interamer. Ciencias Agric., Turrialba, Costa Rica. 66 p.
Verner, J. 1964. Evolution of polygamy in the long-billed marsh wren. Evolution 18: 252-261.
Vite, J. P. 1970. Pest management systems using synthetic pheromones. Contrib. Boyce Thompson Inst. 24: 343-350.
Voûte, A. D. 1964. Harmonious control of forest insects. Int. Rev. Forest Res. 1: 325-383.
Vuilleumier, B. S. 1971. Pleistocene changes in the fauna and flora of South America. Science 173: 771-780.
Vuilleumier, F. 1970. Insular biogeography in continental regions. I. The northern Andes of South America, Amer. Natur. 104: 373-388.
Wadsworth, F. H. 1971. Forestry potential and its development in Central America, p. 171-189. *In* conservación del medio ambiente físico y el desarrollo. Primer Seminario Centro-Americano sobre

el medio Ambiente Físico y el Desarrollo, Antigua, Guatemala. ICAITI/Nat. Acad. Sci.

Wallace, B. 1959. The influence of genetic systems on geographical distribution. Cold Springs Harbor Symp. Quant. Biol. 24: 193-204.

Walter, H., and E. Medina. 1971. Caracterizacion climatica de Venezuela sobre la base de climadiagramas de estaciones particulares. Bol. Soc. Venezolana Ciencias Natur. 24: 211-240.

Watt, K. E. F. 1970. The systems point of view in pest management, p. 71-83. *In* R. L. Rabb and F. E. Guthrie (eds.) Concepts of pest management. Proc. of a Conf. held at North Carolina St. Univ., Raleigh, N. C., 25-27 March, 1970. 242 p.

Wellman, F. L. 1969. More diseases on crops in the tropics than in the temperate zone. Ceiba 14(1): 17-28.

Wellman, F. L. 1972. Tropical american plant disease. Scarecrow Press, Metuchen, N. J. 989 p.

Wells, E. A., and W. H. R. Lumsden. 1970. Trypanosomiasis, p. 309-325. *In* J. W. Davis and R. C. Anderson (eds.) Parasitic diseases of wild mammals. Iowa State University Press, Ames.

Wilkes, H. G., and S. Wilkes. 1972. The green revolution. Environment 14(8): 32-39.

Wilkins, R. M. 1972. Suppression of the shoot borer, *Hypsipyla grandella* Zeller (Lepidoptera: Phycitidae) with controlled release insecticides. Unpublished Ph.D. Thesis, Univ. of Washington, Seattle. 103 p.

Williams, O. B. 1970. Population dynamics of two perennial grasses in Australian semi-arid grassland. J. Ecol. 58: 869-875.

Williams, R. M. C. 1965a. Infestation of *Pinus caribaea* by the termite *Coptotermes niger* Snyder. Proc. 12th. Int. Congr. Entomol. London, 1964. p. 675-676. Academic Press, New York.

Williams, R. M. C. 1965b. Termite infestation of pines in British Honduras. Ministry Overseas Develop., London Overseas Res. Publ. No. 11. 31 p.

Willis, E. O. 1967. The behavior of bicolored antibirds. Univ. Calif. Publ. Zool. 79: 1-127.

Wilson, C. L. 1969. Use of plant pathogens in weed control. Ann. Rev. Phytopathol. 7: 411-434.

Wilson, E. O. 1958. Patchy distribution of ant species in New Guinea rain forests. Psyche 65: 26-38.

Wilson, E. O. 1968. The ergonomics of caste in the social insects. Amer. Natur. 102: 41-66.

Wilson, E. O. 1971. The insect societies. Belknap Press, Cambridge, Mass. 548 p.

Witter, J. F., and D. C. O'Meara. 1970. Brucellosis, p. 249-255. In J. W. Davis, L. H. Karstad, and D.O. Trainer (eds.) Infectious diseases of wild mammals. Iowa State University Press, Ames.

Wolcott, G. N. 1940. An outbreak of the scale insect *Asterolecanium pustulans* Cockerell on Maga, *Montezuma-speciosissima*. Caribbean Forest. 2: 6-7.

Wolcott, G. N. 1946. Factors in the natural resistance of woods to termite attack. Caribbean Forest. 7: 121-134.

Wolcott, G. N. 1954. Termite damage and control as factors in the utilization of timber in the Caribbean area. J. Agric. Univ. Puerto Rico 38: 115-122.

Wolcott, G. N. 1957. Inherent natural resistances of woods to the attack of the West Indian drywood termite, *Cryptotermes brevis* Walker. J. Agr. Univ. Puerto Rico 41: 259-311.

Wolf, L. L., F. R. Hainsworth, and F. G. Stiles. 1972. Energetics of foraging: Rate and efficiency of nectar extraction by hummingbirds. Science 176: 1351-1352.

World Health Organization Expert Committee on Rabies. 1966. Fifth report. World Health Org. Tech. Rep. Ser. 321. 38 p.

Yarbrough, C. G. 1971. The influence of distribution and ecology on the thermoregulation of small birds. Comp. Biochem. Physiol. 39A: 235-266.

Yarwood, C. E. 1968. Tillage and plant diseases. BioScience 18: 27-30.

Zettler, F. W., and T. E. Freeman. 1972. Plant pathogens as biocontrols of aquatic weeds. Ann. Rev. Phytopathol. 10: 455-470.

Section 3

TROPICAL ECOSYSTEM STRUCTURE AND FUNCTION

Ariel Lugo, Team Leader

Mark Brinson
Maximo Cerame Vivas
Clayton Gist
Robert Inger
Carl Jordan
Helmut Lieth
William Milstead
Peter Murphy
Nick Smythe
Sam Snedaker

Contributors: Robert Johannes
William Lewis

I. Introduction

Conventionally, an ecosystem is defined as an integrated set of biological components making up a biotic community plus its abiotic environment (Odum, 1953). The basic biotic components of an ecosystem include (a) producers—the photosynthetic organisms that convert solar energy into chemical energy, (b) consumers that serve as ecosystem regulators, and (c) decomposers that recycle finite material resources. These components, in turn, are composed of species populations. The differences that one observes among ecosystem types can be studied and interpreted in terms of the storages and flows associated with each component, the role that each component assumes in the overall maintenance of that ecosystem and the response of the system to various forcing functions.

Ecosystem study begins first with the delineation of the systems. At the most abstract level these are terrestrial and aquatic ecosystems. Since these two types differ in a number of fundamental ways we have divided the report into two sections, considering each type separately.

Each of these ecosystem types can be subdivided into such groupings as tropical rain forest, tropical dry forest, tropical savanna, desert, and so on. Finally, these types may be divided into the individual stand, lake, or river. Schemes have been devised by Holdridge (1967), Beard (1944), Olson (1970), and many others for classification of tropical ecosystems. In our opinion, further study of classification of ecosystems should be deferred until we understand more about the underlying function of these systems. A classification based on functional criteria probably will be more useful to land managers and planners, as well as ecologists, than one based only on selected features of the producers of the community. Finally, ecosystem studies usually involve an examination of the structure of the system, its function, development, and relationships. In this report, we will focus almost entirely on the first two topics. Development is the subject of Section 4 and relationships between ecosystems are discussed in Sections 5 and 6, since many of the applied ecological problems, such as pollution, involve the interaction between ecosystems.

As defined in this section structure is concerned with the construction of ecosystems, i.e., the variety of species that make up the system inventory, their organization into trophic groups, their biomass, and energy and mineral content. In contrast, function focuses on such processes as energy flow, productivity, mineral cycling, and periodicity.

In each case these structural and functional attributes of ecosystems are often rather poorly established in the tropics so that many more measures are required before we can develop sound principles for prediction. Nevertheless, principles are emerging from the data on tropical systems and these will be discussed separately after we treat the extant data. It is this section that will be most controversial and the material presented here merits special research attention.

II. Recommendations for Research

Ecological research on ecosystem structure and function must be conceived, executed, and interpreted in the context of the needs and realities of society in the tropics, at the same time that it must also be responsive to the further development of ecological theory. We believe that the concept of the ecological system is a useful unifying concept for the design of research questions relevant to both ecological theory and society's needs. This is so because the needs of society are related to the forces acting upon it as part of a system. Thus, ecosystem science helps in the understanding of the complex interactions of societies, as well as the interactions of society with nature, and the functions of nature. In this report we have tried to identify emerging principles of ecosystem science both as they relate to particular ecosystems and to the totality of man's relationship to nature. The actual research questions that emerge from this dual approach are described below.

1. A review of the pertinent literature leads to the hypothesis that process rates (e.g., productivity, decomposition, mineral cycling, succession, etc.) are higher in the tropics than in analogous temperate

regions. The extant data on these rates are not conclusive and sometimes contradictory. This hypothesis needs to be tested in a variety of tropical environments.

2. The significance to society of process-rate data stems from the contributions of ecosystems to man. These services are directly related to the rates of metabolism of the contributing ecosystems. Rates of productivity, decomposition, and cycling must be known, for example, to assess potential organic yields, capacities for waste processing, and water storage alternatives. Rates of succession are required for developing management practices relating to ecosystem recovery, silting of water bodies, and devising new kinds of yield systems.

The challenge is not just to devise ways of increasing the direct benefits to man but also to preserve nature's capacity to provide both direct and indirect benefits in perpetuity. Golley (1972b) summarizes the goal: ". . . to devise systems of land management which will provide food, fiber, recreation, oxygen, water quality, pest control, and the gene pools which mankind will need in the future on a sustained and permanent basis." The gene pools are already disappearing (Gómez-Pompa et al., 1972) and we are therefore obliged to develop an aggressive and creative research effort if our goal is to be achieved.

3. The majority of ecological investigations focus on state variables and relatively short-term (diurnal, seasonal, and annual) biological cycles and processes. With respect to the latter, we have acquired a good working knowledge of the associated periodicities, rhythms, and phenologies, and have come to understand their importance within the ecosystem in which they operate. We know much less about longer term cycles, and we are very ignorant about the mechanisms and possible roles of infrequent or apparently random biological phenomena. Without the requisite understanding, we tend to view such phenomena as destructive and therefore detrimental. These infrequent phenomena are noticeable for their large impacts and tend to be characteristic of either complex "mature" ecosystems or non-steady-state ecosystems with high rates of net production. Some contemporary examples include the following: (a) The coral-eating sea-star *(Acanthaster planci)*, which, until recently, threatened to destroy western Pacific coral reefs that resulted in the formation of an international "control" program. (b) The herbivorous sea urchins that have been decimating relatively large areas of the giant kelp (California) and turtle grass (Florida). (c) The isopod *(Sphaeroma destructor)*, which cuts the prop roots of red mangrove *(Rhizophora mangle)* at the mean-high water line and is apparently destroying the mangrove shore lines in Florida.

Whereas such phenomena are monitored and recorded by the Smithsonian Institution, there is apparently no overt effort being made to fully understand them within the context of the ecosystem. Furthermore, the continuing lack of knowledge generates a self-perpetuating stream of subjective speculations and possibly unneeded control programs.

4. The management of ecosystems requires that some measure of control be exercised over the structure and function of the ecosystem. However, the costs of management decrease if most of the work of maintenance is derived from the natural energy budget. When natural processes are replaced by machines or subsidies, the cost of management increases. For these practical reasons, as well as the theoretical interest in ecosystem control processes, it is important to understand the factors that control ecosystem structure and function.

5. Studies of the dynamics of the important nutrient elements in tropical forest ecosystems should be undertaken. Data on the changes in the cycle that occur during clearing and forest destruction, during cropping, and during abandonment and forest recovery are needed. These studies should include sources and losses of nutrients during all phases, emphasis on the importance of various mechanisms of nutrient cycling such as epiphyllae and mycorrhizae, and observations of changes in the physical properties of soils that affect nutrient availability and the type and quality of vegetation that becomes established in the disturbed areas.

The role of mycorrhizae in mineral cycling probably is of great importance. Much research needs to be carried out on the identification of these organisms, the way in which they transfer nutrients to the roots of tropical trees, which elements are transferred, and the amount of nutrients transferred via mycorrhizae. It will be important to make comparative studies of mycorrhizae in various tropical ecosystems, because the role of mycorrhizae could vary widely from place to place. Also, the consequence of mycorrhizal loss during slash and burn agriculture is of vital importance.

Studies showing that tropical epiphyllae (algae, lichens, bacteria, mosses) are important in fall-out retention (Witkamp, 1970) indicate that these plants also may play a major role in ecosystem nutrition by scavenging nutrients from rainwater. This is a virtually untapped area of tropical research.

6. Several question relating to optimization of ecosystem process are important for ecological theory, as well as sound environmental management. These include determination of optimum rates of ecosystem operation, maintenance of structure and function under stress, and maximization of energy flow per unit area.

7. Traditionally, terrestrial ecosystems have been described in terms of such parameters as species composition and diversity, height, basal areas, presence and frequency of individuals of certain species, age classes, and life forms, qualitative and quantitative structural profiles and maps, biomass, and phenology. For the most part such data are used to describe and to compare differing ecosystem units. Whereas such data are extremely useful in the development of

unifying principles and provide the necessary detail for understanding the whole system, they have not been used to advantage in these regards. We know, for instance, that in the tropics (a) forests and grasslands assume greater heights, (b) there are more epiphytes, (c) there are greater diversities of species, sizes, and life forms, (d) there are more intricate patterns of time and space sharing, (e) there are many diverse mechanisms for nitrogen fixation, and (f) there is generally a greater accumulation of biomass. However, we do not know how each of these characteristics contributes to the long-term survival of tropical ecosystems. Therefore, rather than encourage new field research topics in these areas, we recommend the following procedures for new and continuing tropical research, which may facilitate the development and use of the necessary interpretive framework. These procedural steps are as follows: (a) Design the research such that, in addition to the testing of hypotheses and making comparisons with similar studies displaced in time or space, the results can be interpreted within the context of the next highest level of biological organization. For instance, what is the role of a species in an ecosystem, or how do the components of systems contribute to the systems' energy flow or mineral-cycling networks? (b) Once the research question has been defined, an integrating mechanism should be selected to summarize, interpret, and present the data. Such quantitative techniques include energy-network diagrams, models, or some other technique. Without first defining the integrating mechanism, there is no assurance that all critical factors needed will be considered or that the assembled data can be fully utilized. (c) The current environmental concerns are largely the results of increasing demands being made on finite resources. To ensure that even the most esoteric field research has potential utility, the data should be capable of being expressed in absolute terms, such as quantity per unit area and/or quantity per unit area per unit time. Data expressed in relative terms are usually not capable of interpretation in terms of ecosystems. It should also be apparent to field researchers that the inclusion of these dimensions can be *easily* incorporated into most research schemes. (d) The reviews of the diverse data sets in this report emphasize the overwhelming need for consistent reporting and the use of comparable techniques. The scientific community should work toward an agreement for standardizing the key parameters describing ecosystem structure and function. (e) This review also reveals the propensity for accumulating redundant detail while equally important parameters remain unevaluated. For instance, roots are usually ignored in most studies of forest structure. For those whose job it may be to integrate and synthesize from the extant data, the paucity of key data is especially troublesome.

8. Ecologists have long believed that diversity and stability (or better, resistance to perturbation) are directly related. There are

relatively little data to support this hypothesis. The more diverse the system, the more feedback loops there are and the more the system is resistant to perturbations. A related ecological notion is that energy flows from simple systems to more complex ones, thus as diversity increases, the theory goes, the system becomes more stable and richer.

Understanding the interrelationships of diversity measured by species diversity, stability, and rate of cropping is essential to a theory of ecosystem structure and function. Since the most complex ecosystems—rain forests and coral reefs—are in the tropics, investigations of these interrelationships must include major work in the tropics. This area of research is particularly relevant to man's successful exploitation of ecosystems because successful long-term economic activity is based on predictability and stability.

9. Increased study of regional coupling of ecosystems is necessary since coupling involves the synthesis of theory at a level of organization which is of practical importance to the needs of society. In order to develop a critical level of understanding, we believe that the tools that are applied to this research are almost as important as the research questions themselves. Here we need to apply remote sensing, delineation, and sampling of watershed units, and techniques of modeling to plan research and manipulate data. As discussed earlier, ecological models will be especially useful in this regard since they serve a dual function of planning and are also analytical tools useful in the generation of mechanisms of systems function.

The first priority in a watershed-type study would be to outline the system, diagram its major energy and matter pathways and storages, and then design data collection systems to quantify state variables and flows. However, as discussed in the section on regional coupling, this kind of study demands the understanding of regional implications of life-cycle phenomena, behavioral characteristics of populations, and the seasonality of food availability. This implies coordinated research and regional perspective in data collection and analysis. Some specific areas that need consideration in this context are (a) interaction between the various factors threatening the balance of marine ecosystems, particularly run-off from the land and estuarine degradation; (b) coupling of terrestrial and aquatic ecosystems (this coupling should emphasize the aquatic ecosystems as sinks or indicators of upstream activity, historic and current, and also should compare disturbed and undisturbed ecosystems); (c) the role of marshes and sudds on bordering ecosystems and the change in their nutrient dynamics in agricultural "reclamation," (d) the role of seasonal pulses on metabolism and programming of consumer growth and activity (migration); (e) the effect of land management on estuarine production offshore.

10. Research sites should be selected not by the present distribution of stations but by the data necessary to test general principles of ecosystem structure and function that also are applicable to the needs of man in the tropics. Such strategy would lead to the optimization of data results as they can be utilized at five levels of integration: (a) for understanding the local site, (b) for contribution to the function of the region, (c) for modeling and thus for development of ecosystem theory, (d) for contribution towards the testing of ecological principles, and (e) for direct application to the overriding problem of the balance between man and his life-support system.

III. State of Knowledge

A. Structure of Terrestrial Systems

Several structural features of tropical ecosystems are unique. The tropical rain forest represents the tallest closed vegetation stand on earth with maximum upper canopy heights up to 90 meters, contains the largest biomass and has the greatest variety of species. These characteristics decrease with increasing altitude and decreasing moisture regimes. Excellent summaries of the structural features of tropical biomes are found in Holdridge (1972) and Walter (1971). However, the causes contributing to these characteristic structural properties of the vegetation are largely unknown. They should be thoroughly investigated since management practices are often based on the structural properties of the vegetation.

It is common knowledge that the tropics are more diverse in comparison with the temperate zones, but there are relatively few studies that support this general impression. For example, the Amazon Basin has about 50,000 higher plant species, a number that exceeds the number of higher plants in the entire North American Continent. Species-area curves for a variety of tropical forests show that the number of species increases to the largest plot studied (Fig. 3.1). For most forests, we do not know the size of area required for an adequate sample of the species of trees. We are not aware of similar data for animal species and as far as we know there are no data for the total number of species of plants and animals for any tropical system. We suspect that the diversity of consumers and decomposers follow that of the producers on which they depend, but it is not clear that this is actually so. If it is true, then a measure of the diversity of one part of the system will be adequate to represent the system diversity.

Finally, it also is not clear how species diversity operates in the system. Various ecologists (see Odum, 1971) have suggested that stability of the system and variety of components are directly related, although this supposition needs experimental verification. The implications of diversity have also been discussed in Section 2.

TABLE 3.1. Above-ground Standing Crops in Some Tropical Forests as Reported by Various Authors

Age (yr)	Location	Total Biomass ($g\ m^{-2}$)	Leaf Biomass ($g\ m^{-2}$)	Reference
1	El Verde, P.R.	242 (+ roots)	—	Jordan (1971)
1	Izabal, Guate.	779	654	Popenoe (unpublished)
1	Izabal, Guate.	836	706	Snedaker (1970)
1	Izabal, Guate.	874	734	Tergas (1965)
2	El Verde, P.R.	768 (+ roots)	—	Jordan (1971)
2	Darien, Pan.	1302	362	Golley et al. (1969)
2	Belgian Congo	1323	787	Bartholomew et al. (1953)
2	Izabal, Guate.	1419	953	Snedaker (1970)
2	Guarin, Colom.	1584	260	Ewel in Gamble and Snedaker (1968)
2	Darien, Pan.	2436	296	Golley et al. (1969)
3	El Verde, P.R.	1060 (+ roots)	—	Jordan (1971)
3	Izabal, Guate.	2287	751	Snedaker (1970)
4	Izabal, Guate.	2711	845	Snedaker (1970)
4	Darien, Pan.	3794	594	Golley et al. (1969)
4	Guarin, Colom.	4839	495	Ewel in Gamble and Snedaker (1968)
5	Izabal, Guate.	3667	766	Snedaker (1970)
5	Belgian Congo	7668	563	Bartholomew et al. (1953)

6	Darien, Pan.	4245	655	Golley et al. (1969)
6	Izabal, Guate.	4467	778	Snedaker (1970)
6	Benin, S. Nigeria	4609	419	Nye and Greenland (1960)
7	Izabal, Guate.	4666	1083	Snedaker (1970)
8	Izabal, Guate.	6589	1219	Snedaker (1970)
8	Belgian Congo	12168	538	Nye and Greenland (1960)
9	Izabal, Guate.	7240	1354	Snedaker (1970)
10	Izabal, Guate.	5414	744	Snedaker (1970)
ca. 18	Belgian Congo	12108	644	Bartholomew et al. (1953)
ca. 20	Kumasi, Ghana	9366	446	Nye and Greenland (1960)
40-50	Kade Ghana	26582 (+ twigs)	2025	Greenland and Kowal (1960)
59	El Verde, P.R.	22853	–	Jordan (1971)
60	El Verde, P.R.	23339	–	Jordan (1971)
Mature	Ivory Coast	20000	1000	Ogawa et al. (1965)
Mature	Nigeria	22700	1200	Ogawa et al. (1965)
Mature	Darien, Pan.	26351	794	Golley et al. (1969)
Mature	Thailand	29060	1950	Ogawa et al. (1965)
Mature	Darien, Pan.	37053	1217	Golley et al. (1969)

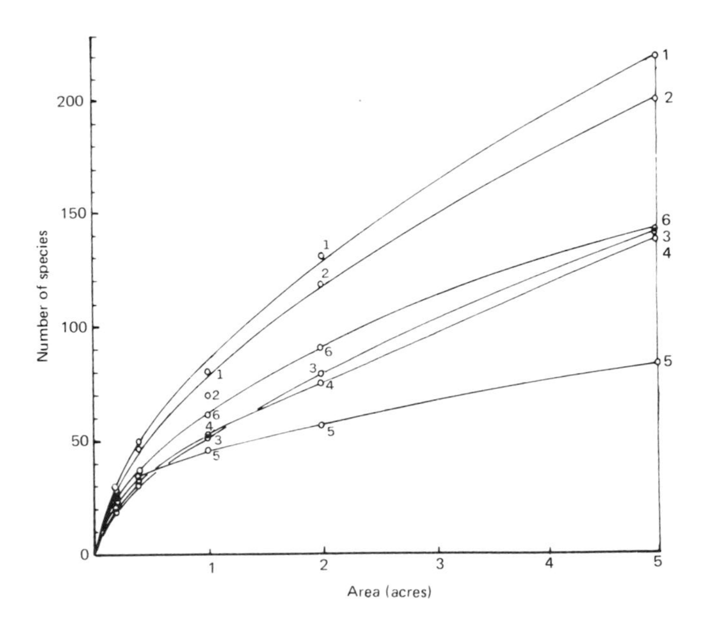

FIG. 3.1. Species-area curves for six sites in Brunei. Adapted from Ross, 1954.

Standing-crop biomass is the quantity of living material in the system at a given time. As mentioned, biomass is often larger in the tropics than in temperate regions. The larger biomass may be due to the fact that in the tropics primary production is higher because of increased insolation, a generally favorable environment, and a longer growing season. The quantity of biomass may also be due to the greater age of undisturbed tropical communities compared with heavily disturbed or youthful temperate systems.

Not only is there a greater accumulation of biomass in the tropics but it apparently reaches a steady-state in a shorter time. The data in Table 3.1 were fitted to a power curve by the least squares regression technique, which yielded

$$\underline{y} = (709.04)\ \underline{x}^{1.0148}$$

where $\underline{y}$ is biomass in g m^{-2} dry weight, $\underline{x}$ is age (years) of vegetation and the correlation coefficient equaled 0.9738.

By solving for $\underline{y}$ (the mean biomass) for various ages ($\underline{x}$), we find that the maximum biomass value for forests is approached in about 30 years at a level of 25 kg m^{-2}. This estimate is supported by studies of Ewel (1968) and Snedaker (1970) for a lowland forest in eastern Guatemala. Temperate forests are thought to reach steady in from 50 to 100 years.

There are very few data on animal biomass in tropical ecosystems (Table 3.2). In forests especially, animals comprise a very small portion of the system mass, yet it must be recalled that most of the above-ground plant mass is wood. Many more data are needed to gain an understanding of the overall role of animals in ecosystem structure and function.

TABLE 3.2. Animal Biomass in Some Tropical Ecosystems

Ecosystem	Animal Biomass (g m-2)	Ratio of Plant/Animal Biomass	References
Tropical forest (Panama)	7.3	4383	Golley et al. (1969)
Tropical forest (Puerto Rico)	12	2264	Odum (1970)
Mangroves (Puerto Rico)	6.4	1762	Golley et al. (1962)
Coral Reefs	143	5	Odum and Odum (1955)
Savanna (East Africa)	45	15	Dasmann (1964)

The energy content of tropical terrestrial ecosystems has been discussed by Golley (1961, 1969). The caloric value of tropical forests is lower than that of forest vegetation in temperate regions, e.g., 4.2 versus 4.6 kcal g^{-1} dry weight of plant stems and Golley suggests that there is a gradient in energy storage in the vegetation with latitude or altitude. However, detailed studies of energy content within a site show significant site and component effects and therefore the energy content should be determined for each individual stand until we have many more data for prediction.

The mineral content of terrestrial ecosystems has been summarized by Rodin and Bazilevich (1967). Calculation of their data for various stands of vegetation (Golley, 1972a) suggests that temperate and tropical communities differ in their elemental composition. Tropical forests contain higher quantities of silica and other elements, but lower quantities of potassium and possibly phosphorus (Fig. 3.2). Again, within- and between-site studies show important differences in mineral content, and many more data are required to understand these differences. For example, data from Panama (Golley et al., in press) suggest that mineral storage in the vegetation increases with the extent of inundation of the stand by water. Riverine forests contained larger quantities of selected elements than adjacent upland forests. Whether this is a conservation or an enrichment phenomenon is unknown.

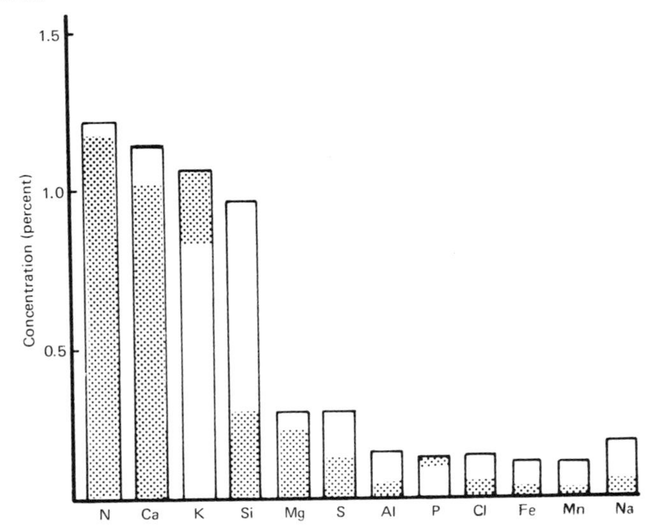

FIG. 3.2. Percent concentration of elements in tropical and temperate vegetation. Open bars denote tropical vegetation; stippled bars denote temperate vegetation. (Data summarized from Rodin and Bazilevich, 1967.)

B. Function of Terrestrial Systems

1. Productivity

Primary productivity is the organic matter input to an ecosystem by the autotrophs during a given period of time. Gross primary production (GPP) is the total conversion through photosynthesis, while net primary productivity (NPP) is the amount of gross production remaining after autotrophic respiration. The organic matter that is not lost from the ecosystem through export or autotrophic respiration becomes available to consumers and decomposers. As Westlake (1963) indicates, ". . . the rate of primary production is ultimately one of the main factors controlling the rate of multiplication and growth of the organisms in a community." It follows that the structural and functional characteristics of an ecosystem are in large part determined by the rate at which the organic matter becomes available. But there is another reason for the strong interest in the levels of NPP attained by various kinds of ecosystems in different climatic regions. Some biologists, such as Westlake (1963), have felt that the rates of NPP attained by natural plant communities may be indicative of rates that can be expected from sophisticated agricultural systems in the same environmental situation. However, energy supplements to agricultural systems have altered this concept.

There are a number of methodological problems in measuring productivity. The types of measurements reported in studies of primary productivity, and the units with which the data are expressed, vary enormously. Units can be converted fairly easily for comparative purposes, although errors in conversion are not infrequent in the literature. As an example, most of the production data reported in the book by Rodin and Bazilevich (1967) are reported as metric centners, a unit unfamiliar to many western scientists. In the English version of the book the centner is defined differently twice; on page 5 as being equivalent to 1 kg and on page 6 as being equivalent to 50 kg. The metric centner is actually equivalent to 100 kg. The point to be made is simply that more uniformity in reporting results would help to eliminate errors in interpretation. The metric system should be used and Westlake's (1963) detailed recommendations concerning the use of units in productivity studies can be followed as a guide.

More important than the selection of units, however, is the need to adequately describe the data that are presented. For example, the time period to which a given value of NPP applies must be stated clearly and the type of drying used to process samples should be specified. It is sometimes difficult to tell whether a given author included below-ground plant parts in his sample, and, if so, how completely they were collected. Total litter fall is sometimes incorrectly used to describe just twigs and leaves, whereas it should include large stems as well.

The methods used to measure NPP vary widely. In herbaceous communities, the harvest method of Boysen-Jansen (1932) is most frequently used, and it can be applied to below-ground as well as above-ground parts. Less desirable is the use of peak standing crop as an approximation of annual NPP; it is generally considered to yield significant underestimates because it does not account for death, herbivory, or export during the growing season. Forest production is often estimated from measurements of stem increment and annual litter production. Sometimes annual litter production alone is used, but it must be multiplied by a factor accounting for trunk and root growth to obtain reasonable estimates of total NPP. Root production is commonly estimated as a fraction of total production equivalent to the fraction of total biomass accounted for by roots. A method of measuring productivity that has recently been employed in some ecosystems, including forests, is the measurement of rates of CO_2 exchange. Odum and Jordan (1970) used a large plastic cylinder enclosing approximately 260 square meters of a lower montane rain forest in Puerto Rico, while Lemon et al. (1970) periodically collected air samples from different vertical positions in a rain forest in Costa

Rica. Such measurements of gas exchange enable rates of ecosystem respiration, gross primary production, and net primary production to be estimated, It is essential that these methods be compared in future studies.

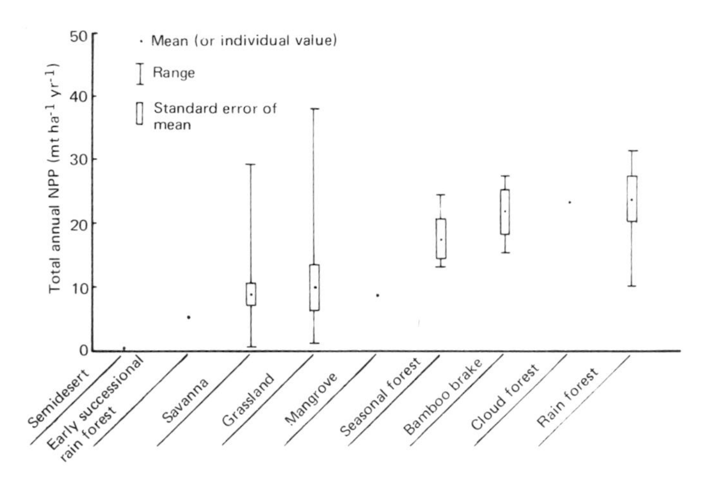

FIG. 3.3. Average annual net primary productivity (NPP) in various tropical ecosystems.

Figure 3.3 contains summary data relating to average levels and variability of NPP in tropical ecosystems. Some of the published data on partial production have been adjusted to total NPP. The variety of terrestrial ecosystems in the tropics exceeds that found anywhere else on earth, ranging from lowland evergreen rain forest to alpine tundra. Consequently, it is meaningless to refer in general terms to the "tremendous" rates of organic production in the tropics—as so many naively optimistic authors have. Usually, references of this sort refer to nonseasonal forests, which in fact occupy less than half of the total land area within tropical latitudes.

Productivity of grasslands varies widely depending in part on the total annual rainfall and the periodicity with which it is distributed. In certain areas of India, for example, a prolonged dry season of nine months greatly restricts the growing season and consequently the total annual NPP even though the total rainfall received is high. As Golley (1972b) suggests, and as would be expected based on evapotranspiration relations, the most productive grasslands appear to be those that occur as successional stages in areas where forest is the expected climax. Of the published data, the maximum site value appears to be 38.1 mt ha^{-1} yr^{-1} measured in a *Heteropogon contortus* dominated grassland in Varanasi, India, which receives in excess of 700 mm of rainfall annually, most of which is distributed over a 3-month period (Ambasht et al., 1972). Based on 11 samples, the average total annual NPP for tropical grassland is estimated to be 10.0 mt ha^{-1} yr^{-1}.

The maximum site value for tropical savanna, excluding irrigated areas, was 29.2 mt ha^{-1} yr^{-1} for a sample in the Congo where rainfall averages 860 mm annually (Bourliere and Hadley, 1970). The minimum site value was 0.7 mt ha^{-1} yr^{-1} for savanna in Chad (Gillet, 1967) and Senegal (Morel and Bourliere, 1962) which receive only about 300 mm of rainfall annually. On the basis of 19 samples the average annual NPP for unirrigated savanna is estimated to be 8.9 mt ha^{-1} yr^{-1}.

The minimum reported value of NPP for the seasonal forest ecosystem is 13.4 mt ha^{-1} yr^{-1}, observed in the Ivory Coast, Africa (Müller and Nielsen, 1965). The maximum site value, measured in deciduous forest patches in the savanna of the Calabozo Plains of Venezuela, is 24.6 mt ha^{-1} yr^{-1} (Medina and Zelwer, 1972). Average NPP, based on only three samples, is 17.8 mt ha^{-1} yr^{-1}.

Annual total NPP of tropical rain forest, based on five samples, averages 23.9 mt ha^{-1} yr^{-1}. The maximum site value is 31.5 mt ha^{-1} yr^{-1} for a forest in the Congo.

Two studies have attempted to measure the integrated metabolism of tropical rain forest. Odum and Jordan (1970) measured rates of CO_2 exchange in a lower montane rain forest in Puerto Rico, and estimated total daily respiration to be 16.4 gC m^{-2}. The ecosystem was considered to be near steady state, and total gross photosynthesis was therefore assumed to be equal to total respiration. NPP was estimated as 12.3 mt ha^{-1} yr^{-1} by subtracting autotrophic respiration from gross photosynthesis. This value agrees well with the allometrically derived value of 10.3 mt ha^{-1} yr^{-1} of Jordan (1971) for the same site. A three-year-old successional rain forest in the same area of Puerto Rico was found to have a total annual NPP of 5.4 mt ha^{-1} (Jordan, 1971). Lemon et al. (1970) measured rates of CO_2 exchange in a 50-year-old rain forest in Costa Rica and found that net ecosystem productivity equaled 3.5 mt ha^{-1} yr^{-1} indicating that the ratio of gross photosynthesis to total respiration in that particular ecosystem was greater than unity.

The production relations of high elevation, montane forests have been little studied in the tropics. The only available production data for evergreen cloud forest, in this case at an elevation of 1,000 m in the Coastal Range of Venezuela, provided an estimated total annual NPP of 23.4 mt ha^{-1} (Medina and Zelwer, 1972).

These data show that the mean levels of productivity range widely, from less than 1 mt ha^{-1} yr^{-1} in deserts to 24 mt ha^{-1} yr^{-1} in evergreen rain forests. The large range of NPP values exhibited by several of the ecosystems reflect differences in the methods used in measuring NPP, as well as the varying effect of environment on this phenomenon. Golley and Lieth (1972), Rosensweig (1968), and others have presented data to predict production from environmental parameters. These predictions need testing in tropical environments.

Tropical ecosystems are more productive than their temperate counterparts. Tropical forests as a whole, with a mean annual NPP of 21.6 mt ha^{-1}, exceed temperate forests, averaging 13 mt ha^{-1}, by a factor of 1.7 and boreal forests, averaging only 8 mt ha^{-1}, by a factor of 2.7. The annual NPP of tropical grassland, averaging 10 mt ha^{-1}, exceeds that of temperate grassland, averaging 5 mt ha^{-1}, by a factor of 2. Estimates of temperate rates of NPP were obtained from Whittaker and Woodwell (1969).

2. Mineral Cycling

The high rate of rainfall in many parts of the tropics, combined with the high rates of litter decomposition and generally old and highly weathered soils results in potentially high rates of nutrient leaching from the ecosystems. Various nutrient conserving mechanisms have been postulated to explain the apparent ability of well-developed tropical ecosystems to avoid leaching losses. One of the most recent suggestions (Went and Stark, 1968) is that mycorrhizae permit the movement of nutrients from decomposing organic material directly to plant roots without moving through the soil. Whatever the mechanism, supposed high nutrient losses in tropical environments lead some ecologists to urge that forests not be removed from these landscapes. This is an extremely important issue since, if true, it is a clear constraint to the utilization of certain lands for forestry and agriculture. Obviously research on mineral cycling is of great theoretical as well as practical interest.

a. Metallic cations. Although mineral-cycling data from the tropics are limited, there are more data on the amounts and flows of the metallic elements than for other aspects of tropical mineral cycling. One reason for this is that the United States Atomic Energy Commission sponsored a large-scale investigation of mineral cycling in Central America and Puerto Rico in an effort to predict the fate of radioactive isotopes should nuclear devices be used to excavate a new transisthmian canal (McGinnis et al., 1969; Templeton et al., 1969; Golley et al., in press). Other reasons for the greater amount of data on cation cycling is the relative simplicity of the metallic cation cycles and the relative ease of analysis for major cations.

An analysis of nutrient cycles in an entire tropical ecosystem by Greenland and Kowal (1960) and Nye (1961) showed that cycling time of most nutrient cations was less than 40 years. A worldwide comparison of amounts of minerals in the biotic portions of ecosystems (Rodin and Bazilevich, 1967) showed that highest levels exist in the wet tropics, and decreasing amounts occur with decreasing available moisture and/or solar radiation on an annual basis.

Jordan and Kline (1972) have compared cycles of major cations in the tropics with cycles in higher latitudes and concluded that the rate of cycling through the biota, but not the mineral soil, is closely correlated with length of growing season. In areas with long growing seasons such as the wet tropics, rates of cycling are highest, and in areas with short seasons, cycling rates are lowest (Table 3.3).

TABLE 3.3. Biotic Compartment Turnover Time (yr) for Calcium as a Function of the Length of the Growing Season

Location	Ecosystem	Length of growing season (day)	Turnover time (yr)
Puerto Rico	Tropical montane rain forest	365	7.5
Ghana	Tropical moist forest	240	8.5
England	Scotch pine plantation	200	10.3
Belgium	Mixed oak	200	13.9
Belgium	Oak-ash	200	22.5
Washington, USA	Douglas fir	200	35.9
USSR	Southern spruce	120	32.5
New Hampshire, USA	Northern hardwoods	120	46.4
USSR	Northern spruce	105	35.1

Data from Jordan and Kline (1972).

b. Carbon, hydrogen, oxygen. Although Deevey (1970) included carbon, hydrogen, and oxygen in his review of worldwide mineral cycles, these three elements are perhaps more suitably studied in other contexts. The carbon cycle is better understood in conjunction with productivity studies, and hydrogen is more logically incorporated into water budgets, although quantification of the water budget is essential to quantification of mineral cycles. A study of the oxygen cycle in relation to terrestrial ecosystems is meaningful only on a worldwide scale. One popular misconception regarding the oxygen cycle is that large numbers of trees in the tropics, especially Amazonia, are needed to generate oxygen for the rest of the world. There will be a net gain of atmospheric oxygen only if the trees that release oxygen and simultaneously bind carbon dioxide do not decompose, but are preserved in bogs or in other anaerobic conditions. There is no reason to believe that such preservation is now occurring, and consequently tropical trees must be using up as much oxygen upon decomposition as they generated during their lifetime.

c. Nitrogen, phosphorus, sulfur. Nitrogen and phosphorus are most often the limiting factors to plant growth in both natural and agricultural ecosystems. However, quantitative estimates of the cycles in natural ecosystems, particularly in tropical ecosystems, are rare.

On a worldwide basis, maximum rates of nitrogen fixation may exceed 200 kg ha^{-1} yr^{-1} in rapidly growing ecosystems (Stewart, 1967). In temperate ecosystems, nitrogen runoff rarely exceeds 12 kg ha^{-1} yr^{-1} (National Academy of Science, 1972), although there may be considerable loss of nitrogen in the systems measured due to harvesting. Since maximum amounts of nitrogen in ecosystems are found in the wet tropics (Rodin and Bazilevich, 1967), rates of nitrogen fixation per unit area per year should be greater in the tropics than in the temperate zones. A greater rate of nitrogen fixation may be a function of the longer growing seasons in the wet tropics. It may also be due to the fact that the great abundance of long-lived leaves in the tropics provides a favorable environment for nitrogen fixation. For example, Edmisten and Kline (1968) found that epiphyllic algae and bacteria were important nitrogen fixers on the leaves of a Puerto Rican rain forest.

In many tropical areas, the high leaching rate of mineral soils results in low levels of phosphorus and sulfur (Olson and Englestad, 1972), and in some areas, sulfur may be a limiting factor (Klinge and Ohle, 1964). However, in relatively undisturbed areas such as the Kade forest in Ghana (Greenland and Kowal, 1960; Nye, 1961), phosphorus may not be limiting, because the phosphorus does not move very deeply into a soil before it is taken up by the vegetation. Luse (1970) showed that uptake of phosphorus from the litter by trees in a Puerto Rican forest was very fast, and little phosphorus was lost into the mineral soil.

3. Phenology

In considering terrestrial tropical climates, we are considering climates in which temperature, photoperiod, and radiation are relatively constant compared with the polar and the temperate zones. Yet there is seasonality in these factors, in addition to water availability, which seems to be the most important factor. In general, the tropical zone is an area in which diurnal temperature variations are at least as great as seasonal variations. Variations in photoperiod and radiation are similarly slight, but all three are important and provide the bases for biological rhythms. Precipitation patterns vary tremendously in the tropics, from nearly uniform throughout the year along the equator to highly seasonal with up to three clearly distinguishable peak seasons. The length of the dry seasons is probably the most significant single factor in determining the ecosystem in areas of adequate total annual rainfall.

These environmental fluctuations influence the reproductive, physiological, and movement patterns of the biota which make up the community. The patterns are best considered under population ecology (Section 2), but those pertaining to the primary producers should be considered here. The processes of special significance to ecosystem function include leaf fall and flowering and fruiting patterns.

a. Leaf production and leaf fall. In the tropical rain forest, the proportion of the whole flora in a given phase of development is approximately the same throughout the year (Richards, 1966). In evergreen plants, the fall of old leaves usually does not have a seasonal periodicity but seems to be only a function of leaf age. In intermittently growing evergreens, leaf shedding often occurs simultaneously with the growth of new leaves or only a few days before (Alvim, 1964). Many trees that exhibit seasonal leaf fall at some distance from the equator become nonseasonal and shed their leaves at regular intervals. Both Fournier (1969) and Alvim (1964) suggest that the breaking of vegetative bud dormancy requires a diurnal temperature variation with relatively low temperatures at night.

In contrast, in the seasonal forest most trees flush in the prerain period (Njoku, 1963). Since a considerable number of upland forest trees put out new leaves well in advance of the rainy season, it seems that deficient soil moisture is not the significant factor. With flushing spread over 11 months and 3 months elapsing between the start of flushing and attainment of maximum activity, it is evident that tropical dry climates exert less control over flushing than the cold winters of the temperate latitudes (Daubenmire, 1972). Day length may be the trigger for flushing. Njoku (1964) found that the onset of dormancy was influenced by day length rather than rainfall. Senescence and abscission are attuned more closely to dryness than day length, with major and minor peaks of these phenomena correlating closely with major and minor peaks of dryness (Daubenmire, 1972). Madge (1965) recorded insignificant leaf fall during the wet season, with pronounced peak during the driest month. Hopkins (1966) also found that maximum leaf fall occurred during the dry seasons. He also found that the timing of these peaks varied somewhat from year to year.

b. Flowering and fruiting. In Panama, Croat (1969) found that only 13 percent of the angiosperms flowered continuously with all others having a seasonal flowering behavior. Over 40 percent of the species had flowering periods as long as 9 months, but the majority had a peak of activity. The peak flowering period for all major components of the vegetation was found to be the dry season. Janzen (1967) has suggested that many tree species in lowland Central America have evolved timing of flowering and fruiting periods within the dry season to maximize vegetative competitive ability in the wet season and maximize the use of pollinators and dispersal agents in the dry season when fewer leaves are present. Snow (1966) concluded that the sequence that *Miconia* species showed in their fruiting peaks had been arrived at through natural selection and reduction of competition for the attraction of birds that disperse the seeds after eating the fruit. Smythe (1970) reached a similar conclusion. He also suggested that synchronous fruiting of larger seeded fruits insured that not all these fruits would be damaged by predators.

Phenological studies have concentrated largely on the periodicity of species populations in response to environment. Research is equally needed on the periodicity of ecosystem processes and on the significance of phenology of species to these processes. The interlinking of species through periodic response is probably of great importance to understanding ecosystem function, yet there are presently few data on this important topic.

4. Role of animals in ecosystems

Animals represent a small fraction of an ecosystem's total biomass and they consume a small fraction of the available live net production of the ecosystem. From this point of view animals appear to play a minor role in energy flow and mineral cycling. However, it is clear that the functional role of animals is in the control of processes such as pollination, fruiting and flowering, litter decomposition, consumption of green plants, as well as productivity and mineral cycling. Examples of the former influences are discussed in Section 2 since they relate to plant-animal or animal-animal interactions. Here we will consider only the impact on nutrients and energy.

Animals affect mineral-cycling rates by shortening the time for minerals in leaves to return to soil. The time for minerals to be released from decomposing leaves depends heavily upon environmental factors; consumption of herbage and excretion in feces represents a shortcut in mineral movement. The direction of flow of nutrients is also affected by animals moving nutrients from their point of accumulation in particular plants. In the case of migratory animals, the nutrients may be carried out of the ecosystems of origin.

The pattern of distribution of nutrients can also be affected by the behavior of animals, which may sequester or aggregate nutrients either for their own nutritional purposes or social functions. Bird and mammal nests and burrows are sites of energy and nutrient aggregations. Nests of social insects, especially termites, represent rich aggregations of mineral nutrients and organic food. Perhaps the most spectacular from the standpoint of extrasomal concentrations of nutrients are the nests and refuse piles of fungus-growing ants and termites, which are confined to the tropics (Weber, 1972; Lugo et al., 1973a; Lee and Wood, 1971).

It has been suggested that animals may have a more homogeneous nutrient concentration than plants since they move and thus sample a variety of nutrient levels (Beyers et al., 1971). In contrast, the individual plants may reflect heterogeneity of the distribution of nutrients in the geological substrate. Very little data exist to test this hypothesis.

In general, the impact of animals on energy flow is similar to their effects on nutrient movement. In forests, herbivorous animals consume usually less than 10 percent of the energy available in primary production. In contrast, in savanna in Africa, where a large mammalian fauna has evolved, Wiegert and Evans (1967) estimate that as much as 50 percent of the energy is consumed by herbivores. Apparently the diverse wild ungulate community of Africa exploits a much wider range of potential plant food than do the herds of domestic cattle in the same region (Talbot and Talbot, 1963). The remaining energy and materials are consumed by decomposer organisms before release to the substrate. Within the consumer or decomposor complex it appears that about 10 percent of the energy consumed is available for transfer to the next level. These estimates all need further study.

C. Fresh-Water Systems: Structure and Function

Historically, aquatic ecosystems have provided the medium in which many important ecological principles concerning production, nutrient cycling, seasonality, population dynamics, diversity, and succession have been tested. Although most of this work has been in temperate regions, there is no *a priori* reason to expect that any of these processes will be different in tropical aquatic ecosystems, except those relating to seasonality. Seasonality in rainfall and humidity may be the key in many cases to the understanding of steady-state processes in these systems.

A disproportionate amount of tropical aquatic research has come from the African continent, no doubt due to interest in the African Rift Lakes resulting from their importance as a regional fisheries resource. Some of the peculiarities of tropical fresh waters discovered in Africa have yet to be clearly demonstrated in the American tropics, but since there is no reason to expect differences between the two continents, examples will be drawn mostly from African studies because of their relative completeness.

Functionally, lakes can be considered as nutrient and sediment traps and rivers as the channels through which the biogeochemical cycle transports its components downhill. Therefore, fresh waters are reflections of the terrestrial ecosystems in their catchment basin.

Specific anomalies, such as low sulfates in many African waters, may imply that the element occurs in low abundances in the terrestrial ecosystems. Large deviations to the other extreme are more likely a result of cities and agriculture. Indeed, lakes and rivers may be our best indicators of the role of man in entire watersheds.

1. Stratification and nutrient cycling in lakes

In lakes deep enough to stratify for longer than brief periods, biologically essential nutrients often become depleted in the surface layers. Nutrient depletion is a combination of two processes: (a) the biological uptake of nutrients that incorporates them into the seston and (b) the progressive sinking of the plankton. The first of these processes can occur within days or weeks of stratification while the second process is slower and progresses throughout the stratification period. Three mechanisms are recognized for regeneration of nutrients to the surface layers: (a) vertical mixing, (b) sinking convective currents, and (c) wind-induced upwelling. These processes are fundamental to other aspects of lake dynamics.

Talling (1969) has reviewed the biological and chemical consequences of vertical mixing in tropical African lakes. Previous to mixing, there are sharp vertical gradients as a result of an impoverished surface layer. Lewis (1973a) in his study of Lake Lanao, Phillipines, found that free nitrate drops to undetectable levels when the water column stabilizes. Dorris (1972) also reported that nitrate reaches detectable levels in the epilimnion only when Lake Atitlán, Guatemala, is freely circulating. Even in deep African lakes which are meromictic (Tanganyika and Malawi) annual vertical mixing is effective in recirculating nutrients to the surface layers. In Lakes Lanao and Atitlán, seasonal circulation is the main mechanism for nutrient distribution, although winds accompanying storms can frequently mix to depths where some nutrients can be regenerated (Lewis, 1973a). Winds may also be important in mixing in Lakes Malawi and Tanganyika.

Talling (1963) also suggests that local cooling due to heat loss to the atmosphere can cause convective currents. The colder and denser waters flow downward along the bottom near the shoreline and thus contribute to the exchange between surface and bottom layers. Talling cited Eccles (1962) as detecting a similar phenomenon in Lake Malawi. This differs from vertical mixing in that there is no direct regeneration of nutrients to surface waters, but an upward displacement of the hypolimnion. It also provides a mechanism for supplying the monimolimnion of meromictic lakes with oxygen. The various mechanisms for regeneration of nutrients merit much fuller study in a greater number of tropical lakes.

Most of the deep lakes just discussed are stratified for longer periods than they mix. As lakes become shallower, the energy requirements to initiate mixing become less, so that very shallow lakes may stratify only irregularly and for very brief periods. Although some shallow tropical lakes are of considerable size, their volumes are small, which amplifies the effect incoming and outgoing water has on their nutrient budget. Lake George (Africa) and Lake Izabal (Guatemala) both have an average residence time of about 1 year even though they are several hundred square kilometers in size (Viner, 1969 and Brinson, 1973). Because of their thorough flushing, the nutrient composition of the incoming water closely resembles that of the lakes. Endorheic lakes, on the other hand, have salinities markedly higher than the waters they receive, and their nutrient budgets should be given special consideration.

Nutrient regeneration to surface waters by whatever mechanism is usually coincident with or closely precedes an increase in phytoplankton production and biomass. For example, the number of phytoplankton, *Melosira*, are synchronized with increased mixing in Lake Malawi (Talling, 1969). Lake Victoria shows the same characteristics (Talling, 1957), and the annual upwelling of the southern area of Lake Tanganyika (Coulter, 1963) is followed by plankton blooms. Weiss (1971b) also observed a maximum of total algae in Lake Amatitlán in November at the time of initial mixing. Increases also occurred in January during destratification and circulation of Lake Atitlán (Weiss, 1971a).

2. Comparison of tropical and temperate lakes

Talling (1965) made a comparative study of Lake Windermere (England) and Lake Victoria (East Africa), which merits close attention as it represents one of the few attempts to carefully analyze differences between temperate and tropical lakes. The annual regimes of water temperature and incident radiation are the principal latitudinal differences. Temperature apparently limits the dominant diatom, *Asterionella* , to the colder waters of Windermere, whereas *Melosira nyassensis* is the dominant phytoplankter of Lake Victoria. During the seasonal diatom maxima of the two lakes, the distribution of chlorophylla and numbers of diatoms differ dramatically with depth but the concentrations in the euphotic zones are similar. Nevertheless, photosynthetic activity per unit volume and per unit area differ greatly. The difference is about 20 times more productivity in Lake Victoria than in Lake Windermere over a year.

Deevey (1957) has also estimated the productivity of Lake Atitlán to be of the same order as the maximum observed in temperate lakes (~ 5.14 g O_2 m^{-2} day^{-1}). Amatitlán has increased severalfold in productivity in the last 20 years and is 50 to 100 times (net production) greater than Atitlán (Weiss, 1971a), illustrating the diversity of lakes in geographically similar areas as well as the effect of cultural eutrophication.

The role of nutrients in temperate and tropical freshwater may differ since Deevey (1957) has noted that the nitrogen-phosphorus ratios (N/P) were lower in Central American lakes than in temperate waters by a factor of as great as five. He suggested that nitrogen deficiency may be general in the tropics. These relationships imply that eutrophication problems also may vary in temperate and tropical waters. Since eutrophication can result from an increase in the absolute concentration of N and P and shift in the N/P ratio can amplify the process, we might expect in tropical lakes that an addition of sewage could not only increase the concentration of N and P but also change their ratio. This supposition is not supported by study of Lake Amatitlán where the N increased by a factor of six but phosphorus remained constant (Weiss, 1971b). Even so, there has been a four- to sevenfold increase in production in Lake Amatitlán. Clearly these nutrient processes require detailed study under tropical conditions.

Brylinsky and Mann (1973) used multiple regression and factorial analysis for data collected from widely ranging latitudinal zones as part of an International Biological Program study. When all latitudes were considered, they found that solar energy input had a greater influence on production than nutrient concentration. However, for lakes within a narrow range of latitude, nutrient-related variables became more important in influencing primary production. These comparative studies are extremely important since they suggest the limits to community process we can expect under normal environmental conditions. In this sense they form a baseline against which change in the aquatic system can be measured. Clearly many more studies of this type are needed.

3. River systems

The principal river systems of the tropics, including the prominent Amazon, Congo, Nile, and Mekong Rivers, are of immense unrealized biological significance. Due to their close relationship with the economic growth of the regions they occupy, these rivers will undoubtedly be changed markedly within the next few decades. Whatever information is lost in the transition of these systems to a postindustrial state will be an unfortunate handicap to tropical ecological investigation and may well hinder effective management of the associated resources.

The volume and drainage area of the major tropical rivers is large and forms a habitat complex unlike any temperate counterpart. The flow of the Amazon is eight times and the Congo two times that of the Mississippi (Sioli, 1964). The mainstem of these river systems, although impressive in volume and biotic diversity, constitutes only a small portion of the environmental complex derived from runoff in tropical regions. Tributary rivers and streams determine water chemistry of the mainstem and support a significant biota of their own. Such habitats are ecologically unknown except in certain restricted regions in South Africa and the Amazon. Valuable but scattered studies of the smaller rivers and streams are cited in a review by Hynes (1970).

Some unique communities are associated with tropical rivers and the relative importance of various community types may differ greatly from the temperate zone. One striking example is provided by the enormous mats of floating vegetation or "floating meadows" (Marlier, 1962; Sioli, 1968; Junk, 1970). Backwaters and waterbodies within the flood plain such as the várzea lakes of the Amazon (Sioli, 1968; Schmidt, 1969) are also important environmental components of tropical river systems, especially since the floodplain has frequently not been altered by man. Direct economic importance of the backwaters and várzea lakes is probably great because they may supply the nursery requirements for commercially important fishes (W. Junk, personal communication).

Few biological or physicochemical generalizations can be made regarding lotic environments of the tropics. This is due partly to ignorance and partly to the great range of conditions prevailing in these environments. The humid tropics contribute the vast majority of flow, but many major rivers traverse or receive seasonal contributions from arid regions. Very dilute rivers are frequently found in the lowland humid tropics. The Orinoco, for example, carries only about 50 ppm TDS (Livingston, 1963). Low dissolved-solids content of many lowland tropical rivers is due in part to efficient absorption and retention of salts by organic material in the watersheds. Parent material and relief also account for wide variation in tributary composition in the largest river systems. Sioli (1963, 1964) distinguishes the "white" water of the lower Amazon, which receives its sediment load from areas of high relief, from both the "clear" waters such as the Rio Tapajos and the "black" waters such as the Rio Negro, which flow from areas of low relief. The black waters of the Rio Negro are acid (pH 3.8 to 4.9), low in electrolytes (Ca $<$ 5 mg/l, Mg $<$ 0.4 mg/l, HCO_3 $<$ 0.04 m val/l) and high in humic colloids and humic acid, which give the brown color to the water and reduce the light needed for photosynthesis. Biological production in the water is therefore very low (Sioli, 1954, 1955, 1963, 1965a, b). Janzen (unpublished, 1973) finds evidence that the humic materials may also be toxic to some organisms.

Sioli (1963, 1964) has also recognized the role of bedrock in producing runoff of varied composition within the Amazon basin. As would be expected, specific regions of the tropics have peculiarities that are reflected in their rivers and streams. The waters of East Africa, for example, are remarkably low in sulfate (Beauchamp, 1953). The great range of variation in African surface waters is documented by the work of Talling and Talling (1965). Higher level of silicates in tropical freshwaters, predicted by Hutchinson (1957) on the basis of early data and the higher mobility of silicate under alkaline conditions, has recently been verified by Talling and Talling (1965) in Africa and Kobayashi (1966) and Lewis (1973b) in Southeast Asia. They demonstrated the presence of high silicate concentrations even when other plant nutrients were scarce. Community composition and adaptive features can obviously be affected by such factors. More generalizations of this type can be expected due to the fundamental differences between tropical and temperate weathering processes.

Terrestrial studies have shown that in the lowland tropics standing crop provides large nutrient pools due to the low ratios of productivity to biomass of the mature tropical terrestrial ecosystems. As the mean age of these ecosystems is rapidly lowered by man, one might reasonably expect a radical change in chemical composition of the more dilute tropical waters and a concurrent change in biota.

Variations in the biota of rivers can be largely understood in terms of chemical composition, rate of flow, and turbidity. Most tropical rivers, like those of the temperate zone, can be expected to show seasonal variation in volume of flow, hence seasonal variation in chemical composition and turbidity. Kobayashi's (1959) study of river chemistry in Thailand has demonstrated tropical seasonality particularly well, and Prowse and Talling (1958) have documented seasonal variation of the Nile plankton.

Other climatic and biotic factors influence tropical rivers but their effect may be more difficult to judge *a priori*. Tropical rivers are distinct from temperate-zone rivers in relative seasonal uniformity of temperature and insolation, although mean temperatures can be consistently very low at high altitudes. Biological functions and interactions may well be affected by this uniformity in ways that are unique to the tropics.

Zoogeographical accidents are another principal source of contrast between temperate and tropical running water. The rivers of North America, for example, are dominated by cyprinid, catastomid, ictalurid, and to a lesser extent centrarchid and percid fishes. The Amazon, by contrast, supports a high diversity of characin and siluriform fishes that contribute to a total fauna of nearly 2,000 species (Darlington, 1957). Large African rivers are comparably diverse in fish fauna, but have distinctive dominant groups such as the mormyrid fish. Whether the organization of food webs in tropical rivers is radically different due to the special adaptive capabilities of the top trophic levels in the various major river systems is not known. The origin of and ecological support for high diversity within the top trophic levels of tropical rivers also await explanation.

The large tropical river systems are without doubt the greatest habitat frontier in modern aquatic ecology. Study of these systems would not only increase our general knowledge of freshwater ecology, both in a descriptive and a conceptual sense, but would provide important data for planning a rational pattern of resource use.

4. Management of freshwater systems

The extent to which aquatic primary productivity can be converted into products useful to mankind poses an important management question. Most data show that tropical waters produce more fish than temperate waters (Table 3.4) and that fish-pond culture has the highest yields in tropical latitudes (Table 3.5). Responses to fertilization of ponds are also greater in the tropics. Research on the basic mechanisms underlying these differences is needed.

TABLE 3.4. Fish Yields Reported for a Diversity of Lake Sizes and Latitudinal Locations

Waterbody	Fish crop per unit area ($kg\ ha^{-1}$)
Swiss and German alpine lakes	13
Eastern Germany, freshwater	21
Lake Mendota, Wisconsin	22
Lake George, Uganda	117
Lake Nakivali, Uganda	188
Lake Kitangiri, Uganda	316
Ponds fed by hot springs, Trogong, Java	2,000-10,000

Data from Hickling (1966).

TABLE 3.5. Fish Production in Fertilized and Unfertilized Ponds

Area	Unfertilized ($kg\ ha^{-1}$)	Fertilized ($kg\ ha^{-1}$)
Northern Europe and Siberia	70-80	200-300
Southern Europe	220-250	600-700
Israel	90-100	300-400
Malacca, Malaysia	314	1681

Data from Hickling (1966).

By impounding rivers, man has been able to create "lakes" in months or years compared with the millennia that are normally required for the formation of natural lakes. Only organisms preadapted to the new environment can participate in community organization.

Thus, the establishment of reservoirs represents a kind of ecological succession. To ignore reservoirs would be to disregard potentially important experiments that could lead to a better understanding of succession in aquatic ecosystems. On a more practical level, it should be possible to predict the characteristics and change on the basis of past examples and histories.

Observations of reservoir succession reported in the literature often reveal disquieting problems. These initially unstable ecosystems may show explosive growths of floating plants, increases in human pests and disease, and downstream effects on fisheries as in the case of the Nile dams' effect on fisheries offshore of the Nile delta (George, 1972).

From our knowledge of succession and some of its underlying principles, a newly formed reservoir might be considered to be a heterotrophic system with an initially high respiration resulting from a high organic load. The temporary initial success of fisheries often observed may be one of the expressions of this unbalance or non-steady-state condition. Even the thermal regime requires time to stabilize so that a regular pattern of circulation is attained and planktonic and benthic communities are established.

Flooding of large areas of land typically results in deoxygenation of the overlying water. This was especially marked in Brokopondo, Surinam, where the forest was not cleared and dead submerged trees restricted vertical circulation (Leentvaar, 1966). In Lake Volta, Africa (Ewer, 1966), low oxygen concentrations were partially attributed to water turbidity, which limited photosynthesis. Several years were required before phytoplankton became abundant in more than the

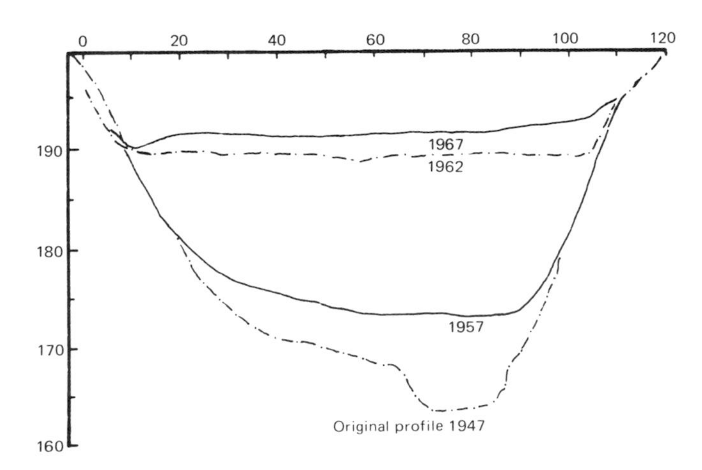

FIG. 3.4. Progressive deposition of sediments in Anchicaya Reservoir. Adapted from Allen, (1972).

upper few meters (Biswas, 1972). Lake Kariba (Harding, 1966) followed a similar trend and showed a progressive increase in the oxygenated water mass. The increase in oxygen is probably more a result of a decrease in oxygen-consuming organic matter submerged during the original flooding.

In impoundments such as Lake Kariba and Lake Volta, which take several years to fill to capacity (Worthington, 1966), it is unlikely that sedimentation will significantly reduce the capacity of the reservoirs. Smaller reservoirs that fill with water in a few months are becoming quite common for hydropower generation in mountainous areas of Central and South America. In the case of the Anchicayá Reservoir in Colombia (Allen, 1972), mass deposits of sediment seriously reduced its storage capacity to a small fraction of the original volume, apparently because the watershed was not adequately protected from clearing and road building (Fig. 3.4).

Theoretically it should be possible to predict reservoir behavior based on such factors as the morphology of the basin, local climate, residence time of the water, and water quality. Such predictions would be an important contribution for future hydrological projects and might lead to a more realistic and rational approach to their development.

D. Coastal Marine Ecosystems

Coastal marine ecosystems comprise a variety of types that may differ greatly from those in temperate regions. Beaches, which occur on shorelines where the substrate is loose and in transport, are comprised of sediment that is sorted according to particle size by the mechanical action of the waves. Consequently, the particle size varies depending upon its position on the beach. This process acts as an environmental determinant affecting the zonation of organisms on the beach. As an ecological unit, beaches boast the most predictable and most stereotyped of all systems. Tropical beaches have faunal and floral assemblages almost similar to those of subtropical and temperate beaches, even having the same genera (Gonzalez-Liboy, 1971; Wade, 1967; Pearse et al., 1942). There are no great differences in the assemblages of beaches on both the Pacific and the Caribbean sides of Central America (Dexter, 1972) when beaches of the same particle size ranges are considered. Major differences exist in the inhabitants of cobble as opposed to sand beaches. Tidal amplitude may also affect the nature and zonation of beaches.

Rocky shores are characterized by fairly vigorous mechanical pounding of the waves, a broad swash zone, and a broad spray zone. As is the case with beaches, a rocky shore as an ecological unit boasts no major single feature that will significantly distinguish it from a rocky shore of higher latitude except that its species have a tropical distribution.

In marked contrast to beach and rock shore habitats, the coral-reef ecosystem is unique to the tropics. It is one of the most highly productive and diverse tropical systems. Goreau (1961) states that coral reefs are also the single most extensive shallow-reef community in the world. Thus, on the basis of area, inherent productivity, or unique structure, coral reefs merit further intensified study by ecologists.

The coral-reef ecosystem has been seldom studied metabolically. The most detailed investigation (Johannes et al., 1972) is of reef communities at Eniwetok Atoll. This study considered the flux of nitrogen, phosphorus, carbon, and energy of the entire reef system and showed that rates of phosphorus recycling and nitrogen were unusually efficient. Interestingly, the study also showed that the algal flat community associated with the reef may be equally or even more productive than the coral reef itself. There are a number of studies of reef structure, including Wells (1957), Storr (1694), Wiens (1962), and Yonge (1966), and of reef history, and the response of reefs to human disturbance.

Besides their productivity and diversity (Odum and Odum, 1955), coral reefs are of considerable direct value to man. In many regions, they are a source of high-quality protein. There are many poisonous coral species, and these may have considerable pharmacological importance (Baslow, 1969). It is well known that coral reefs are a wave barrier protecting islands and coastlines. They are also of considerable significance as a tourist attraction. Tropical-island communities frequently depend upon beaches of coral sand, lagoons behind reefs, and skin diving on reefs as the basis for important tourist income.

Clearly, it will be important to understand this resource in order to conserve and utilize it effectively. However, management of reefs will involve new information and techniques. For example, managers of temperate fisheries usually handle one or two species. On coral reefs, over 150 species of fish and invertebrates may be harvested for protein. Management requires information on the growth rate and life history of key species. We also know very little about the growth rate of any coral-reef fish or invertebrate. Indeed, systems as complex as coral reefs have seldom been effectively studied as systems, except in a very few instances. Further, we know little about the impact of pollution and sedimentation on coral reefs. Johannes (1973) states that sedimentation probably causes more reef destruction than any other stress. However, he also finds that the ability of coral to withstand stress varies with the nature of the sediment, the type of reef, and its environment. Clearly, the stability of the reef system requires detailed and thorough investigation.

Mangrove ecosystems, with adjacent concentrations of *Thallasia* beds, are usually found in areas of low coastlines or in areas where coral reefs serve as buffers to high-wave action. The environment in which mangroves grow is nutrient rich, with the fertility depending on the volume of terrestrial runoff. The physiognomy, productivity, and standing biomass of mangroves may be characterized as tall stands with high rates of productivity and high levels of biomass characteristic of areas of nutrient rich and high-volume water runoff. With decreasing runoff, mangrove stands become smaller and less productive. The zonation of mangrove ecosystems, their horizontal and vertical stratification, adaptations to the salinity gradients, and metabolism have been studied and summarized in the literature by numerous authors, including Davis (1940), McNae (1966), Walter (1971), and Thom (1967).

Several recent studies have stressed the importance of mangrove detritus as a basis for estuarine production (Heald, 1969; Odum, W. E., 1970). In Florida, these authors estimate that 800 g. dry matter of mangrove leaves are produced annually on a square meter. Less than 5 percent of this production is consumed directly, and less than 1 or 2 percent is stored as peat. About one-half of the production is consumed by mangrove bacteria, fungi, and detritus feeders, and one-half is flushed to shallow coastal embayments. This input of mangrove to coastal waters may well be crucial to maintenance of invertebrate and fish production and should be studied in more detail. We especially need to study the utilization of this detritus through food chains.

Besides their role in production of material for off-shore fisheries mangroves also stabilize sediment, protect coastlines, provide a source of fuel, construction material, and tannin, and other uses. In fact, in parts of the tropics mangrove forests are managed intensively for wood and fuel production. Conversion of mangrove to rice land and to other uses requires a knowledge of the role of mangrove in the landscape and more research is needed before such conversion schemes can be adequately evaluated.

E. Modeling Tropical Systems

Modeling is a logical method of studying the structure and function of ecological systems. However, there are few models in the literature dealing with tropical systems. Three models have been developed to deal with consequences of radioactive contaminations in tropical forests. Golley et al. (in press) discuss stable mineral cycling in Panamanian wet forests located in Darien Province, Panama. Jordan and Kline (1972) discuss models of ^{90}Sr in a Puerto Rican rain forest, and Odum et al. (1970) present models of tritium dynamics in this same forest. Jordan et al. (1972) have applied modeling techniques to understanding the relative stability of mineral cycling. Other mineral cycling models include those of Greenland and Kowal (1960) and Nye (1961) for forests in Ghana and Stark (1971) for the Amazonian forest. An energy model for a tropical rain forest was published by Odum (1970) In his book (Odum, 1971), and in other papers (Odum,

1967a, b; Lugo, 1970) models of savannas, shifting cultivation, and several other tropical systems (Fig. 3.5) are presented with energy values for all flows and storages. However, only a few of these models have been simulated, as were the mineral cycling models discussed above. Exceptions are the models of mangrove ecosystems (Lugo et al., 1974, in press), models of terrestrial microcosms (Odum et al., 1970), and the potassium model of tropical moist forest (Golley, 1972a).

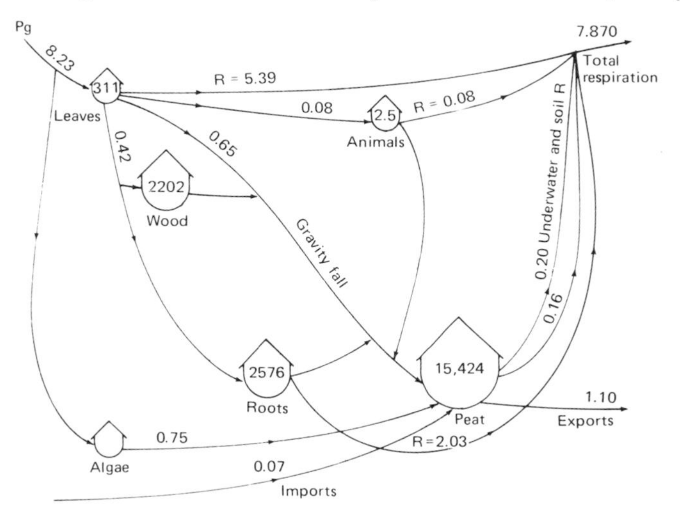

FIG. 3.5. Carbon budget for a Puerto Rican red mangrove forest. (Adapted from the data of Golley, Odum, and Wilson 1962.) Storages are in gC m^{-2} and flows are in gC m^{-2} day^{-1}.

Major problems in the modeling of tropical systems include the lack of data on key components such as decomposers. Although as much as 90 to 95 percent of the energy passes through this component, relatively little is known about its structure or function. Further, the high diversity of tropical systems is difficult to simulate, yet is clearly of central significance in understanding the dynamics of these systems. Forcings on the system are poorly understood since there has been little experimental work on whole systems in the tropics, with the notable exception of the radiation experiment in Puerto Rico (Odum, 1971).

IV. Emerging Principles

A. Primary Productivity

Annual levels of tropical net primary productivity on an area basis appear to exceed temperate levels by 2 to 4 times for similar terrestrial ecosystems. The differences in productivity between temperate and tropical forests is apparently due largely to production of leaf biomass since wood production is not necessarily greater in the tropics (Jordan, 1971). Tropical productivity appears to be limited most often by moisture although nutrient levels are also limiting in some sites, as is temperature at higher elevations.

B. Mineral Cycling

If many tropical ecosystems do in fact have relatively high rates of net primary productivity, then it may be inferred that rates of mineral turnover are also relatively high. For example, tropical forests, especially in the lowlands, typically have thin litter layers. Confirming data on actual rates of decomposition, however, are lacking. It may be that the decomposition surface in some tropical ecosystems, especially forests, is more heterogeneous than in temperate ecosystems. As an example, termite mounds may be considered sites of concentrated decomposition. Many tropical ecosystems appear to have nutrient conserving mechanisms that enable storage of essential nutrients. These mechanisms might include storage in long-lived biotic elements such as stems and roots, mycorrhizal linkage between decomposing organic matter and roots, uptake of nutrients from leaves before leaf fall, and other factors. The need for further study in all of these areas is strongly indicated.

C. Predictability of the Environment

For many years it was widely assumed that the conditions that support life never fall below an optimal level in neotropical forests. Many theories regarding evolution, selection, and species diversity were based upon this assumption. For the assumption to be valid the environmental conditions must be at or near an ideal level. Throughout the greater part of the area covered by tropical forests there is variation in the amount of monthly precipitation. Much of this area is subjected to one or two dry seasons during a "normal" year. It is becoming clear that in many tropical forests resources available to the local organisms vary considerably from one time of the year to another (Smythe, 1970). In fact, some recent studies (Lloyd et al., 1968) show that even in certain rain-forest areas that have no dry season, flooding of streams with the accompanying flushing of the forest by heavy rains occurs at random intervals over a year. Other studies (Snow, 1966) tend to indicate that medium term (1 to 3 years) resource predictability may also be less than assumed.

Lack of environmental predictability on a short- or medium-term basis may be a key to the maintenance or evolution of a high diversity of species (Lloyd et al., 1968; Slobodkin and Sanders, 1969). The fact that we know so little of stability, seasonal variation, and predictability in the neotropical forest, when viewed in the light of the further fact that this biome is disappearing at an almost incredible rate, makes this a primary research goal for the immediate future. Long-term studies should begin immediately. They are tedious and unromantic but the knowledge gained both for heuristic and applicable purposes, would repay considerable effort.

D. Environmental and Internal Controls of Ecosystems

Compositional, structural, and dynamic features of ecosystems seem in many cases clearly controlled by abiotic forces. The correlation of ecosystem responses to some of the more obvious environmental sources and forces are at present accepted textbook knowledge; many less obvious ones, however, are insufficiently known. In future research it will be useful to order the environmental parameters in a matrix; so that the limiting nature of each factor can be specified.

Internal controls of tropical ecosystems are unique in many respects and must be studied in much more detail. The influence of primary producers on the stand environment is best known but is insufficient for prediction. The role of consumer organisms is also poorly known at the ecosystem level. Further, all of predator-prey, parasite-host, and animal-plant interactions need study from the ecosystem as well as from the population point of view.

E. Optimization and Maximization

Lotka (1956) and others suggest that natural selection operates to maximize energy flow in individuals. Energy flow is hence maximized in the assemblage of organisms in an ecosystem. Structural and functional features of the ecosystem are related to the overall flow of energy through the system. This includes the efficiency of energy capture through photosynthesis and its subsequent use through respiration. Within the bounds of water and nutrient availability, succession may occur and result in the development of more structure which in turn appears to augment the efficiency of the system to transform solar radiation into organic matter. Ecosystem development leveling off at an upper limit of productivity reflects increasing respiration rates associated with increasing structural complexity in the system. Data on productivity and energy flow of ecosystems suggest that overall energy flow is optimized at that point in which a given structural complexity maximizes its amplifying potential in terms of increased solar energy converted.

It appears that in ecosystems where environmental conditions are optimum for life the organism biomass is highest and the energy cost of maintenance lowest, and in ecosystems where environmental conditions are harsh biomass is lowest and the maintenance cost highest. Further study of the ratios between production, respiration, and biomass will clarify these concepts.

F. Regional Coupling Mechanisms

The ecosystems discussed here are coupled to other systems to produce regional landscape units. There are many examples of coupling mechanisms; but relatively little is known of their specific nature and research. Many couplings involve the timing of the life cycle or behavior of a population through allocthonous inputs to the system. Couplings also involve net movement of nutrients and/or organic material either towards the sea (downhill) or towards the land (uphill). There is the tendency for most biological processes to conserve nutrients in the terrestrial systems (uphill transport), whereas physical factors predominate in the downhill transport. Ecosystems with net organic export (mangroves and agriculture) also have a net nutrient input. Rivers and sewage plants receive organic inputs and generally export inorganic materials.

Some examples of couplings are as follows:

1. Drift lines occur after a moderately heavy surf over shallow tropical waters. Many beaches accumulate mounds of marine grasses that get uprooted or broken by the mechanical action of the waves. These drift-line accumulations, "arribazones," provide considerable biomass to what could be an otherwise depauperate beach. The grass and/or algal mounds decompose on shore while providing a haven not only for amphipods, decapods, and isopods, but also for spiders, insects, and lizards who belong to terrestrial food chains.

2. Salt-water intrusion into streams and rivers in the shape of wedges over the bottom are well-documented coupling situations. The upstream ("uphill") limit of the salt-water wedge is determined by river runoff and by tide. Wedges provide an excellent way for marine organisms to exploit fresh-water resources. In the same water-column parcel, one may find fresh-water organisms above and marine forms below.

3. Some animals have complex life cycles in which the larval (or nymphal) habitat is radically different from the adult couple habitats via its life cycle. In the tropics certain terrestrial insects, such as scarab beetles, move from the subterranean to arboreal strata in great numbers. Aquatic insects, frogs, and their aquatic larvae move materials between land and fresh-water. These may be viewed as lateral movements within a habitat or ecosystem. Riverine fishes in the tropics commonly move into small tributaries after flooding to spawn, thus constituting upstream motion within an ecosystem or movement between ecosystems.

4. Catadromous and anadromous fishes are probably the most spectacular examples of biological coupling between marine and fresh-water ecosystems. Many fishes, including some commercially important species such as mullets, pass their juvenile stages in estuaries and their adult stages in open coastal waters. Larval stages or juve-

niles of many marine invertebrates exploit mangrove estuaries. Large tropical rivers are well known for their many typical fishes, such as sea bass, that move far upstream.

5. Marine birds come to the mangrove to defecate and the result is an accumulation of nutrients in the system. The algae *Halimeda* and *Udotea* accumulate calcium carbonate, which upon their death provides the algal marl favorable to mangrove colonization and shoreland development. Sea cucumbers in the mangrove channels pass fecal pellets that stabilize silts.

The occurrence of these regional couplings have relevance to socioeconomic phenomena. Man, his action, and his dependency on natural phenomena are all related to the ability of regions to maintain a livable environment. This not only means clean air, soil, and water, but it also includes an environment where agriculture, forestry, and urbanization are possible at low costs to society. Man does not have enough energy resources to subsidize a completely artificial environment; therefore, he needs natural systems to maintain regional homeostasis. For this we must understand and protect those regional couplings of ecosystems that are essential for striking the proper balance between the population of man and those natural systems.

G. Modeling and Systems Analysis

Sensitivity analysis is a technique to determine which component of a system is most sensitive to change. The results of a sensitivity analysis of a relatively simple Puerto Rican rain-forest mineral-cycling model (Jordan and Kline, 1972) supported the statement of Cooper (1969) that "hypothetical simple systems can be far more sensitive to changes in the organizational structure of the relations between components than to changes in the values of the components themselves."

An ecosystem can be unstable, bounded, or asymptotically stable. When ecosystem processes are all stable, they can be compared by calculating their stability (Jordan et al., 1972). In a stable mineral-cycling system, it is the ratio of input of an element to its recycling that determines the relative stability of the cycling system. If recycling is large relative to input, the recovery time from a perturbation is relatively long and the system has low relative stability. If little recycling takes place, recovery time is short and relative stability is high.

Relatively large storage compartments, such as the wood in a tropical rain forest, buffer the cycles of minerals against variations in the physical environment (Burns, 1970). Small compartments and high turnover rates such as may exist in the decomposer portion of a system result in a high sensitivity of the components to changes (Jordan and Kline, 1972).

Because loss of elements from ecosystems occur primarily from leaching of the soil, it was hypothesized that the tightness of a cation cycle in an ecosystem could be predicted by the cations replacing power (Marshall, 1964) on soil surfaces. A test of the hypothesis (Jordan and Kline, 1972) showed that for mono- and divalent cations in a Puerto Rican rain forest, relative replacing power was a good indicator of the relative tightness of an element's cycle. Modeling efforts are limited both by the lack of data from tropical systems and techniques for modeling itself. Both aspects merit further research.

References

Allen, R. N. 1972. The Anchicayá hydroelectric project in Colombia: Design and sedimentation problems, p. 318-342. *In* M. T Farvar and J. P. Milton (eds.) The careless technology. Ecology and international development. Natural History Press, Garden City, New York.

Alvim, P. de T. 1964. Tree growth periodicity in tropical climates, p. 479-495. *In* M. H. Zimmerman (ed.) The formation of wood in forest trees. Academic Press, New York.

Ambasht, R. S., A. N. Maurya, and U. N. Singh. 1972. Primary production and turnover in certain protected grasslands of Varanasi, India, p. 43-50. *In* P. M. Golley and F. B. Golley (compilers) Tropical ecology with an emphasis on organic production. University of Georgia, Athens.

Bartholomew, W. V., J. Meyer, and H. Laudelont. 1953. Mineral nutrient immobilization under forest and grass fallow in the Yangambi (Belgian Congo) region. Publ. Inst. Nat. Etude Agron. Congo Belge, Ser. Sci. No. 57. 27 p.

Baslow, M. H. 1969. Marine pharmacology. Williams and Wilkins Co., Baltimore, Maryland. 286 p.

Beard, J. S. 1944. Climax vegetation in tropical America. Ecology 25: 127-158.

Beauchamp, R. S. A. 1953. Sulphates in African island waters. Nature 171: 769.

Beyers, R. J., M. H. Smith, J. B. Gentry, and L. L. Ramsey. 1971. Standing crops of elements and atomic ratios in a small mammal community. Acta Theriologica 16: 8-18.

Biswas, S. 1972. Ecology of phytoplankton of the Volta Lake. Hydrobiologia 39: 277-288.

Bourliere, F., and M. Hadley. 1970. The ecology of tropical savannas. Ann. Rev. Ecol. Syst. 1: 125-152.

Boysen-Jensen, P. 1932. Die Stoffproduktion der pflanzen. Gustav-Fisher, Jena. 108 p.

Brinson, M. M. 1973. The carbon budget and energy flow of a tropical

lowland aquatic ecosystem. Unpublished Ph.D. Thesis, Univeristy of Florida, Gainesville.
Brylinsky, M., and K. H. Mann. 1973. An analysis of factors governing productivity in lakes and reservoirs. Limnol. Oceanogr. 18: 1-14.
Burns, L. A. 1970. Analog simulation of a rain forest with high-low pass filters and a programatic spring pulse, p. I-284-I-289. *In* H. T. Odum and R. F. Pigeon (eds.) A tropical rain forest. A study of irradiation and ecology at El Verde, Puerto Rico. U.S. Atomic Energy Commission, Oak Ridge, Tennessee.
Cooper, C. F. 1969. Ecosystems models in watershed management, p. 309-324. *In* G. M. van Dyne (ed.) The ecosystem concept in natural resource management. Academic Press, New York.
Coulter, G. W. 1963. Hydrological changes in relation to biological production in southern Lake Tanganyika. Limnol. Oceanogr. 8: 463-471.
Croat, T. B. 1969. Seasonal flowering behavior in central Panama. Ann. Missouri Bot. Garden 56: 295-307.
Darlington, P. J. 1957. Zoogeography: The geographical distribution of animals. Wiley, New York. 675 p.
Dasmann, R. F. 1964. African game ranching. Pergamon Press, Oxford, England. 75 p.
Daubenmire, R. 1972. Phenology and other characteristics of tropical semideciduous forest in N.W. Costa Rica. J. Ecol. 60: 147-170.
Davis, J. H. 1940. The ecology and geologic role of mangroves in Florida. Papers from Tortugas Lab. 32: 305-410.
Deevey, E. S. 1957. Limnologic studies in middle America with a chapter on Aztec limnology. Trans. Conn. Acad. Arts Sci. 39: 213-328.
Deevey, E. S. 1970. Mineral cycles. Sci. Amer. 223: 149-158.
Dexter, D. M. 1972. Comparison of the community structure in a Pacific and an Atlantic Panamanian sandy beach. Bull. Mar. Sci. 22: 449-462.
Dorris, T. C. 1972. La ecologıa y la pesca del Lago de Atitlán. El Centro de Investigaciones de Embalses de la Univ. Estatal de Oklahoma, E. U. Publicación Especial, 1.
Eccles, D. H. 1962. An internal wave in Lake Nyasa and its probable significance in the nutrient cycle. Nature 194: 832-833.
Edmisten, J. A., and J. R. Kline. 1968. Nitrogen fixation by epiphyllae, p. 141-143. *In* The rain forest project annual report. Puerto Rico Nuclear Center Report 119. Rio Piedras, Puerto Rico.
Ewel, J. J. 1968. Dynamics of litter accumulation under forest succession in Eastern Guatemalan lowlands. Unpublished M. S. Thesis, University of Florida, Gainesville.
Ewer, D. W. 1966. Biological investigations on the Volta Lake, May 1964 to May 1965, p. 21-30. *In* R. H. Lowe-McConnell (ed.) Man-made lakes. Symp. Inst. Biol. London 15. Academic Press, London.

Fournier, L. A. 1969. Observaciones preliminares sobre la variación altitudinal en el numero de famílias de arboles y de arbustos en la vertiente del Pacífico de Costa Rica. Turrialba 19: 548-552.

Gamble, J. F., and S. C. Snedaker. 1968. Final report. Agricultural ecology. Battelle Mem. Inst., Columbus Lab. and Univ. of Florida. Center for Trop. Agr.

George, C. J. 1972. The role of the Aswan High Dam in changing the fisheries on the Southeastern Mediterranean, p. 159-178. *In* M. T. Farvar and J. P. Milton (eds.) The careless technology. Natural History Press, Garden City, New York.

Gillet, H. 1967. Essai d'evaluation de la biomase vegetale en zona sahelienne (vegetation annuelle). J. Agr. Trop. Bot. Appl. 14: 123-258.

Golley, F. B. 1961. Energy values of ecological materials. Ecology 42: 581-584.

Golley, F. B. 1969. Caloric value of wet tropical forest vegetation. Ecology 50: 517-519.

Golley, F. B. 1972a. The potassium cycle in a tropical moist forest ecosystem p. 349-358. *In* Memorias de Symposia. 1st. Congr. Latinoamericano y Mexico de Bot. Sociedad Botanica de Mexico, S. C., Mexico, D. F.

Golley, F. B. 1972b. Summary, p. 407-413. *In* P. M. Golley and F. B. Golley (compilers) Tropical ecology with an emphasis on organic production. University of Georgia, Athens.

Golley, F. B., H. T. Odum, and R. F. Wilson. 1962. The structure and metabolism of a Puerto Rican red mangrove forest in May. Ecology 43: 9-19.

Golley, F. B., J. T. McGinnis, and R. G. Clements. 1969. Final report, terrestrial ecology. Battelle Memorial Instit., I.O.C.S. Mem. BMI-26. 70 p.

Golley, F. B., and H. Lieth. 1972. Bases of organic production in the tropics, p. 1-26. *In* P. M. Golley and F. B. Golley (compilers) Tropical ecology, with emphasis on organic production. University of Georgia, Athens.

Golley, F. B., J. T. McGinnis, R. G. Clements, G. F. Child, and M. J. Duever. 1974. Mineral cycling in a tropical moist forest ecosystem. University of Georgia Press, Athens.

Gómez-Pompa, A., C. Vázquez-Yanes, and S. Guevara. 1972. The tropical rain forest: A nonrenewable resource. Science 177: 762-765.

Gonzales-Liboy, J. H. 1971. The zonation and distribution of marine beach macroinvertebrates at Plaza Mani, Mayaguez, Puerto Rico. Unpublished M.S. Thesis, Dept. Mar. Sci., Univ. Puerto Rico, Mayaguez.

Goreau, T. F. 1961. Problems of growth and calcium deposition in reef corals. Endeavor 20: 32-40.

Greenland, D. J., and J. M. L. Kowal. 1960. Nutrient content of the moist tropical forest of Ghana. Plant and Soil 12: 154-174.

Harding, D. 1966. Lake Kariba: The hydrology and development of the fisheries, p. 7-19. *In* R. H. Lowe-McConnell (ed.) Man-made lakes. Symp. Inst. Biol. (London) 15. Academic Press, London.

Heald, E. J. 1969. The production of organic detritus in a south Florida estuary. Unpublished Ph.D. Thesis, Univ. Miami, Fla.

Hickling, C. F. 1966. Fish culture. Faber and Faber, London.

Holdridge, L. R. 1967. Life zone ecology. Trop. Sci. Center, San José, Costa Rica. 260 p.

Holdridge, L. R. 1972. Forest environments in tropical life zones: A pilot study. Pergamon Press, New York 746 p.

Hopkins, B. 1966. Vegetation of the Olokemeji forest reserve; Nigeria IV. The litter and soil with special reference to their seasonal changes. J. Ecol. 54: 687-703.

Hutchinson, G. E. 1957. A treatise on limnology. Vol. 1. Wiley, New York. 1015 p.

Hynes, H. B. N. 1970. The ecology of running waters. Liverpool University Press, Liverpool, 555 p.

Johannes, R. E. 1973. Coral reefs and pollution, p. 364-371. *In* Marine pollution and sea life. FAO Conference on Marine Pollution. Fishing News (Books), London.

Johannes, R. E. et al. 1972. The metabolism of some coral reef communities: A team study of nutrient and energy flux at Eniwetok. BioScience 22: 541-543.

Janzen, D. H. 1967. Synchronization of sexual reproduction of trees within the dry season in Central America. Evolution 21: 620-637.

Janzen, D. H. 1973. Tropical black-water rivers, animals, and mast fruiting by the Dipterocarpaceae, unpublished.

Jordan, C. F. 1971. Productivity of a tropical rain forest and its relation to a world pattern of energy storage. J. Ecol. 59: 127-142.

Jordan, C. F., and J. R. Kline. 1972. Mineral cycling: Some basic concepts and their application in a tropical rain forest. Ann. Rev. Ecol. Syst. 3: 33-50.

Jordan, C. F., J. R. Kline, and D. S. Sasscer. 1972. Relative stability of mineral cycles in forest ecosystems. Amer. Natur. 106: 237-253.

Junk, W. 1970. Investigations on the ecology and production-biology of the "floating meadows" *(Paspalo-Echinocholetum)* on the middle Amazon. Amazoniana 2: 449-495.

Klinge, H., and W. Ohle. 1964. Chemical properties of rivers in the Amazonian area in relation to soil conditions. Verh. Int. Verein. Limnol. 15: 1067-1076.

Kobayashi, J. 1959. Chemical investigations on river waters of south eastern Asiatic countries. I. The quality of the waters of Thailand. Berichte Ohara Inst. Land. Biol. 11: 167-233.

Kobayashi, J. 1966. Silica in fresh water and estuaries, p. 41-55. *In* H. L. Golterman and R. S. Clymo (eds.) Chemical environment in the aquatic habitat. NV Noord-Hollandsche Uitgevers Maatschappij, Amsterdam.

Lee, R. E., and T. G. Wood. 1971. Termites and soil. Academic Press, New York. 251 p.

Leentvaar, P. 1966. The Brokopondo research project, p. 33-42. *In* R. H. Lowe-McConnell (ed.) Man-made lakes. Symp. Inst. Biol. (London) 15. Academic Press, London.

Lemon, E. R., L. H. Allen, and L. Muller. 1970. Carbon dioxide exchange of a tropical rain forest. Part II. BioScience 20: 1054-1059.

Lewis, W. M. Jr. 1973a. The thermal regime of Lake Lanao (Philippines) and its theoretical implications for tropical lakes. Limnol. Oceanogr. 18: 200-217.

Lewis, W. M. Jr. 1973b. The thermal regime, chemistry and phytoplankton ecology of Lake Lanao, Philippines. Unpublished Ph.D. Thesis. Indiana University, Bloomington. 264 p.

Livingston, D. A. 1963. Chemical composition of rivers and lakes. Data of geochemistry, 6th ed. Chap. G., U. S. Geol. Surv., U.S. Gov't Printing Office, Washington, D.C.

Lloyd, M., R. F. Inger, and F. W. King. 1968. On the diversity of reptile and amphibian species in a Bornean rain forest. Amer. Natur. 102: 497-515.

Lotka, A. J. 1956. Elements of mathematical biology. Dover, New York. 465 p.

Lugo, A. E. 1970. Energy flow in some tropical ecosystems. Soil Crop Sci. Soc. Florida 29: 254-264.

Lugo, A. E., E. G. Farnworth, D. J. Pool, P. Jerez, and G. Kaufman. 1973a. The impact of the leaf cutter ant *Atta colombica* on the energy flow of a tropical wet forest. Ecology 54: 1292-1301.

Lugo, A. E., M. Sell, and S. C. Snedaker. 1974. Mangrove ecosystem analysis. *In* B. C. Patten (ed.) System analysis and simulation ecology. Vol. III. Academic Press, New York, in press.

Luse, R. A. 1970. The phosphorus cycle in a tropical rain forest, p. H-161-H-166. *In* H. T. Odum and R. F. Pigeon (eds.) A tropical rain forest. A study of irradiation and ecology at El Verde, Puerto Rico. U.S. Atomic Energy Commission, Oak Ridge, Tennessee.

McGinnis, J. T., F. B. Golley, R. G. Clements, G. I. Child, and M. J. Duever. 1969. Elemental and hydrologic budgets of the Panamanian tropical moist forest. BioScience 19: 697-700.

McNae, W. 1966. Mangroves in eastern and southern Australia. Austr. J. Bot. 14: 67-104.

Madge, D. S. 1965. Leaf fall and litter disappearance in a tropical forest. Pedobiologia 5: 277-288.

Marlier, G. 1962. Etude sur les lacs de l' Amazonie centrale. Cadern. da Amazonia, Manaus-Amazonas 5: 1-51.

Marshall, C. E. 1964. The physical chemistry and mineralogy of soils. Vol. 1. Wiley, New York. 388 p.

Medina, E., and M. Zelwer. 1972. Soil respiration in tropical plant communities, p. 245-267. In P. M. Golley and F. B. Golley (compilers) Tropical ecology with an emphasis on organic production. University of Georgia, Athens.

Morel, G., and F. Bourliere. 1962. Relations ecologiques des avifaunes sedentaire et migratrice dans une savane sahelienne du bas senegal. Terre Vie 4: 371-393.

Müller, D., and J. Nielsen. 1965. Production brute, pertes par respiration et production nette dans la foret ombrophile tropicale. Forstlige Forsogsvasen Danmark 29: 69-160.

National Academy of Sciences. 1972. Accumulation of nitrate. Comm. on Nitrate Accumulation, Agr. Board, Div. Biol. and Agr., Nat. Res. Council, Nat. Acad. Sci., Washington, D.C.

Njoku, E. 1963. Seasonal periodicity in the growth and develpoment of some forest trees in Nigeria. J. Ecol. 51: 617-624.

Njoku, E. 1964. Observation on seedlings. J. Ecol. 52: 19-26.

Nye, P. H. 1961. Organic matter and nutrient cycles under moist tropical forest. Plant and Soil 13: 333-346.

Nye, P. H., and D. J. Greenland. 1960. The soil under shifting cultivation Commonw. Bur. Soils, Harpenden. Tech. Comm. No. 51. Commonwealth Agr. Bur., England. 156 p.

Odum, E. P. 1953. Fundamentals of ecology. Saunders, Philadelphia. 384 p.

Odum, H. T. 1967a. Energetics of world food production, p. 55-94. *In* Vol. III. The World Food Problem. Report, President's Sci. Advisory Comm., The White House.

Odum, H. T. 1967b. Work circuits and systems stress, p. 81 to 138. *In* Symposium on primary productivity and mineral cycling in natural ecosystems. University of Miami Press, Miami, Florida.

Odum, H. T. 1970. Summary: An emerging view of the ecological system at El Verde, p. I-191-I-281. *In* H. T. Odum and R. F. Pigeon (eds.) A tropical rain forest. A study of irradiation and ecology at El Verde, Puerto Rico. U.S. Atomic Energy Commission, Oak Ridge, Tennessee.

Odum, H. T. 1971. Environment, power, and society. Wiley, New York. 331 p.

Odum, H. T., and C. F. Jordan. 1970. Metabolism and evapotranspiration of the lower forest in a giant plastic cylinder, p. I-165-I-189. *In* H. T. Odum and R. F. Pigeon (eds.) A tropical rain forest. A study of irradiation and ecology at El Verde, Puerto Rico. U.S. Atomic Energy Commission, Oak Ridge, Tennessee.

Odum, H. T., A. Lugo, and L. Burns. 1970. Metablism of forest floor microcosms, p. I-35-I-56. *In* H. T. Odum and R. F. Pigeon (eds.) A tropical rain forest. A study of irradiation and ecology at El Verde, Puerto Rico. U.S. Atomic Energy Commission, Oak Ridge, Tennessee.

Odum, H. T., and E. P. Odum. 1955. Trophic structure and productivity of a windward coral reef community on Eniwetok Atoll. Ecol. Monogr. 25: 292-320.

Odum, W. E. 1970. Pathways of energy flow in a south Florida estuary. Unpublished Ph.D. Thesis, University of Miami, Miami, Florida.

Ogawa, H., K. Yoda, K. Ogino, and T. Kira. 1965. Comparative ecological studies on three main types of forest vegetation in Thailand. II. Plant biomass. Nature and Life in S.E. Asia 4: 49-80.

Olson, J. S. 1970. Geographic index of world ecosystems, p. 297-304. *In* D. E. Reichle (ed.) Analysis of temperate forest ecosystems. Springer-Verlag, New York.

Olson, R. A., and O. P. Engelstad. 1972. Soil phosphorus and sulfur, p. 82-101. *In* Soils of the humid tropics. Nat. Acad. Sci., Washington, D. C.

Pearse, A. S., H. S. Humm, and G. W. Wharton. 1942. Ecology of sand beaches at Beaufort, N. C. Ecol. Mongr. 12: 136-190.

Prowse, G. A., and J . F. Talling. 1958. The seasonal growth and succession of plankton algae in the White Nile. Limnol. Oceanogr. 3: 222-238.

Richards, P. W. 1966. The tropical rain forest. Cambridge Univ. Press, Cambridge. 450 p.

Rodin, L. E., and N. I. Bazilevich. 1967. Production and mineral cycling in terrestrial vegetation. Oliver and Boyd, Edinburgh. 288 p.

Rosensweig, M. L. 1968. Net primary productivity of terrestrial communities: Prediction from climatological data. Amer. Natur. 102: 67-74.

Ross, R. 1954. Ecological studies on the rain forest of southern Nigeria. III. Secondary succession in the Shasha forest reserve. J. Ecol. 42: 259-282.

Schmidt, G. W. 1969. Vertical distribution of bacteria and algae in a tropical lake. Int. Revue Ges. Hydrobiol. 64: 791-797.

Sioli, H. 1954. Gewässerchemie und Vorgänge in den Böden im Amazonasgebiet. Naturwiss 41: 456-457.

Sioli, H. 1955. Beiträge zur regionalen Limnologie des Amazonasgebietes. III. Üeber einige Gewässer des oberen Rio-Gebietes. Arch. Hydrobiol. 50: 1-32.

Sioli, H. 1963. Beiträge zur regionalen Limnologie des brasilianischen Amazonasgebietes. V. Die Gewässer der Karbonstreifen Unteramazoniens. Arch. Hydrobiol. 59: 311-350.

Sioli, H. 1964. General features of the limnology of Amazonia. Verh. Inter. Verein. Limnol. 15: 1053-1958.

Sioli, H. 1965a. A limnologia e a sua importância em pesquisas da Amazonia. Amazoniana 1: 11-35.

Sioli, H. 1965b. Bermerkungen zur Typologie Amazonischer Flüsse. Amazoniana 1: 74-83.
Sioli, H. 1968. Principal biotopes of primary production in the waters of Amazonia, p. 591-600. *In* R. Misra and B. Gopal (eds.) Proc. Symp. Recent Ad. Trop. Ecol. Part II. International Society for Tropical Ecology, Varanasi, India.
Slobodkin, L. B., and H. L. Sanders. 1969. On the contribution of environmental predictability to species diversity, p. 82-95. *In* Diversity and stability in ecological systems. Brookhaven Symp. Biol. 22. Brookhaven National Laboratories, Upton, New York.
Smythe, N. 1970. Relationships between fruiting seasons and seed dispersal methods in a neotropical forest. Amer. Natur. 104: 25-35.
Snedaker, S. C. 1970. Ecological studies on tropical moist forest succession in eastern lowland Guatemala. Unpublished Ph.D. Thesis, University of Florida, Gainesville. 131 p.
Snow, D. W. 1966. A possible selective factor in the evolution of fruiting seasons in a tropical forest. Oikos 15: 274-281.
Stark, N. 1971. Nutrient cycling: I. Nutrient distribution in some Amazonian soils. Trop. Biol. 12: 24-50.
Stewart, W. O. P. 1967. Nitrogen fixing plants. Science 158: 1426-1432.
Storr, J. F. 1964. Ecology and oceanography of the coral reef tract, Abaco Island, Bahamas. Geol. Soc. Amer. Spec. Publ. 79.
Talbot, L. M., and M. H. Talbot. 1963. The high biomass of wild ungulates on east African Savannah. Trans. 28th. N. Amer. Wildlife and Natur. Resource Conf. p. 465-476 Wildlife Management Institute, Washington, D.C.
Talling, J. F. 1957. Some observations on the stratification of Lake Victoria. Limnol. Oceanogr. 2: 213-221.
Talling, J. F. 1963. Origin of stratification in an African rift lake. Limnol. Oceanogr. 8: 68-78.
Talling, J. F. 1965. Comparative problems of phytoplankton production and photosynthetic productivity in a tropical and a temperate lake, p. 399-425. *In* C. R. Goldman (ed.) Primary productivity in aquatic environments. Mem. Ist. Ital. Idrobiol, 18 suppl. University of California Press, Berkeley.
Talling, J. F. 1969. The incidence of vertical mixing, and some biological and chemical consequences in tropical African lakes. Verh. Inter. Verein. Limnol. 17: 998-1012.
Talling, J. F., and J. B. Talling. 1965. The chemical composition of African lake waters. Int. Rev. Ges. Hydrobiol. 60: 421-463.
Templeton, W. L., J. M. Dean, D. G. Watson, and L. A. Rancitelli. 1969. Freshwater ecological studies in Panama and Colombia. BioScience 19: 804-808.
Tergas, L. E. 1965. Correlation of nutrient availability in soil and

uptake by natural vegetation in the humid tropics. Unpublished M. S. Thesis, University of Florida, Gainesville. 64 p.

Thom, B. G. 1967. Mangrove ecology and deltaic geomorphology. J. Ecol. 55: 301-343.

Viner, A. B. 1969. The chemistry of the water of Lake George, Uganda. Verh. Inter. Verein. Limnol. 17: 289-296.

Wade, B. 1967. Studies on the biology of the West Indian beach clam, *Donax denticulatus* L. I. Ecology. Bull. Mar. Sci. 17: 149-174.

Walter, H. 1971. Ecology of tropical and sub-tropical vegetation. Oliver and Boyd, Edinburgh.

Weber, N. A. 1972. Gardening ants. The attines. Mem. Amer. Phil. Soc. 92: 1-146.

Weiss, C. M. 1971a. Water quality investigations, Guatemala: Lake Atitlán 1968-1970. Dept. Envir. Sci. Eng., Univ. North Carolina, Chapel Hill, Regional School Sanitary Eng., Univ. San Carlos, and Inst. Geogr. Nac., Guatemala, May, 1971; Chapel Hill, N. C., ESE Publ. 274. 175 p.

Weiss, C. M. 1971b. Water quality investigations, Guatemala: Lake Amatitlán, 1969-1970. Dept. Envir. Sci. Eng., Univ. North Carolina; Regional School Sanitary Eng., Univ. San Carlos; and Inst. Geogr. Nac., Guatemala, May, 1971. Chapel Hill, N. C. ESE Publ. 281.

Wells, J. W. 1957. Coral reefs. Mem. Geol. Soc. Amer. 67: 609-631.

Went, F .W., and N. Stark. 1968. Mycorrhizae. BioScience 18: 1035-1039.

Westlake, D. F. 1963. Comparisons of plant productivity. Biol. Rev. 38: 385-425.

Whittaker, R. H., and G. M. Woodwell. 1969. Measurement of net primary production of forests. Brookhaven Nat. Lab. Rep. (BNL 14056), Brookhaven National Laboratories, Upton, New York.

Wiegert, R. G., and F. C. Evans. 1967. Investigations of secondary productivity in grasslands, p. 499-518. *In* K. Petrusewicz (ed.) Secondary productivity of terrestrial ecosystems. Polish Acad. Sci., Warsaw.

Wiens, H. J. 1962. Atoll environment and ecology. Yale University Press, New Haven, Conn. 532 p.

Witkamp, M. 1970. Mineral retention by epiphyllic organisms, p. H-177-H-179. *In* H. T. Odum and R. F. Pigeon (eds.) A tropical rain forest. A study of irradiation and ecology at El Verde, Puerto Rico. U.S. Atomic Energy Commission, Oak Ridge, Tennessee.

Worthington, S. B. 1966. Introductory survey, p. 3-6. *In* R. H. Lowe-McConnell (ed.) Man-made lakes. Symp. Inst. Biol. (London) 15. Academic Press, London.

Yonge, C. M. 1966. Introduction to the Great Barrier Reef. Austr. Nat. Hist. 15: 233-236.

Section 4

RECOVERY OF TROPICAL ECOSYSTEMS

Arturo Gómez-Pompa, Team Leader

Ana Luisa Anaya
Frank Golley
Gary Hartshorn
Daniel Janzen
Martin Kellman
Lorin Nevling
Javier Peñalosa
Paul Richards
Carlos Vázquez
Paul Zinke

Contributor: Sergio Guevara

I. Introduction

The vegetation on the earth has existed for millions of years but has evidenced continuous change over time. For example, the angiosperms, originating in the Jurassic or possibly even Triassic time (Smith, 1973) about 181 million years ago, partially replaced the previous plants that had dominated the vegetation. Presumably these plants, with their animal life, formed ecological communities under the same processes currently operating on the earth. The contemporary communities in turn are the result of continuous plant and animal evolution under the selection pressure of geological, biotic, atmospheric, and hydrobiological processes.

Superimposed upon these continuing long-term events are also short-term processes on a scale of one to several hundred years. These processes are termed ecological successions. Succession may be the result of severe disturbance associated with natural catastrophes, man-induced changes, or may be due to very subtle processes. For example, the sporadic presence of an herbivore species eliminating a few individuals of a tree species may have a strong influence on the community; a fire sweeping through the forest floor may change the numbers and kinds of organisms living in the system. In each instance, the individual species in the ecosystem interact in such a way that a disturbance to one of them or to one part of the system may affect the entire community.

Secondary succession is a general term describing the changes of ecosystems following the incomplete destruction of a community. Secondary succession thus defined can occur in a small patch of "untouched" forest following the fall of a tree or over hundreds of hectares of abandoned farm land. There can be no doubt that the proportion of the earth's vegetation in some stage of secondary succession has been steadily increasing due to manipulation of the vegetation accompanying the growth of the human population.

Clearly, the current trends of population growth and increased land use in tropical America will expand the areas occupied by managed and secondary ecosystems so that the primary ones will become the exception (UNESCO, 1960) as has already occurred in many areas of the Old World tropics. For this reason it is of the utmost importance to understand the ecological processes of communities undisturbed by man to be able to compare the impact of man's actions against a standard. Further, knowledge of successional dynamics may allow man to manipulate and control successional processes and speed the recovery of land damaged through misuse.

Secondary succession has been studied mainly in temperate areas and knowledge of the tropical process is derived from studies in only a few localities (Budowski, 1961, 1963; Ewel, 1971a; Kellman, 1970a; Gómez-Pompa et al., 1963, 1964; Kenoyer, 1929; Ross, 1954; Rico, 1972; Sarukhán, 1964; Sousa, 1964). Therefore the amount of relevant and reliable data is very limited. For example, many studies restrict themselves to lists of the species found in secondary communities. While these data are useful, few generalizations can be derived from them.

One of the most obvious problems that a scientist has to deal with when working in tropical communities is the determination of its history, which together with the determination of the age of a community and its future behavior are central problems of ecology. These studies can be extremely difficult because secondary communities are often dominant in the areas to be studied. The general trend to study only primary vegetation ("potential vegetation") in vegetation surveys suggests that the successional series are well understood, which is far from the truth in most cases (Gómez-Pompa and Cázares, 1970).

Studies of secondary succession will continue to be important in understanding basic biological problems. The behavior of populations during succession has raised very important questions in relation to organic evolution (Gómez-Pompa, 1971a) in the tropical environment. Succession involves interactions between all community processes such as mineral cycling, organic productivity, rates of decomposition, microorganismal activities, and biotic interactions of all kinds. These changes occur within a framework of a changing physical environment. Moreover, the similarities and differences between succession in tropical and temperate systems are poorly understood. Further, localities under different tropical conditions (wet-dry; high-low elevations; different latitudes) should be compared to understand the variability of the process within the region.

Beside these purely ecological problems, elucidation of the properties of secondary communities is also urgently needed to develop a rational system to exploit tropical resources. Since more and more land is being converted to secondary systems, it is important to understand productivity changes through time as compared with steady-state systems (Gómez-Pompa et al., 1964). Recent interest has been directed to the tropics as sources of germ plasm. Tropical natural areas have great plant and animal species richness and are natural gene and genotype banks, which are rapidly disappearing. From this point of view, it has been suggested (Gómez-Pompa et al., 1972) that the course of natural regeneration of tropical rain forests is basically different from that of temperate forests, due to differences in the biological interactions and in the viability and dormancy of seeds of primary forest trees. There is a much higher danger of species loss in the tropics, which might have serious repercussions on mankind. Research on secondary succession must be intensified in order to produce land-use alternatives allowing conservation of these gene pools for the future.

This last point may be sufficient justification for the importance of secondary succession studies but there are many additional reasons. Basic facts from secondary succession will undoubtedly assist the development of neotropical forestry (Foggie, 1960). As discussed in Section 3, the tropical forest has the largest standing-crop type of community on the earth. Maintenance of a potential forest ecosystem in a managed grassland or shrub community (i.e., an early succession community) requires a great input of energy to arrest the process of succession. The maintenance of the balance of nutrients, biotic interactions, and energy input to stabilize managed ecosystems is a tremendous challenge (The Institute of Ecology, 1972). The side effects of artificial nutrient and energy input and the effects of biocides on natural succession may be severe and unexpected. There is a great need to apply ecological principles to the analysis of productivity of all kinds of systems, not only with a view to obtaining the highest usable yields of organic net productivity per area, but also to maintain sustained productivity for long periods of time. There is much to learn from recovery systems in connection with biological control of pests and crop diseases, competition, and efficiency of productivity in managed communities.

We can expect that man will continue to make errors in land-use management since our knowledge is incomplete and our ability to use this knowledge is imperfect; thus we will be dependent on natural or on managed recovery processes to reestablish vegetation and animal life, to reduce erosion, and to provide alternative forms of production to maintain original ecosystems for their aesthetic values. Our concern now is to improve our ability to predict what will happen to a drastically changed ecosystem such as defoliated forest in Vietnam or areas in large tropical deforestation programs.

II. Recommendations for Research

1. Establishment of permanent centers (which may be existing stations) to study succession over the long time periods is necessary in this area of research.

2. Establishment of replicated plots to study short- and long-term effects of disturbance in several major types of tropical communities is needed. Microclimate, soils, animals, and plants should be monitored as they change through time. There should be adequate replication of plots to allow repeated disturbance of the same and different kinds and to allow intermittent destructive sampling. Among the disturbance types should be those that deliberately mimic current and anticipated man-induced disturbances. Whenever possible, plot treatments and characterization should be coordinated and standardized from region to region. Adjacent areas of undisturbed vegetation which can act as controls to the experiments should be conserved as well.

3. Regional studies of secondary communities should be undertaken with preference given to areas where past history is accurately known and the data should be recorded so that it will be possible to generate testable hypotheses about the processes of change, to make the descriptions relevant to understanding rapidly changing habitats of importance to man, and to encourage other investigators to conduct their studies in the same or similar areas.

4. "Ecological life histories" (autecological population studies) of organisms that are prominent in and/or have a conspicuous effect on disturbed habitats are needed. Examples of appropriate organisms are rice rats *(Oryzmys)*, ticks *(Ixodidae)*, kiskadees *(Pitangus sulphuratus)*, chrysomelid beetles (Chrysomelidae), bacteria, ants, vines, and species of *Piper*, *Eupatorium*, *Cecropia*, *Bauhinia*, or *Mimosa*. A given ecological life history should extend across as wide a range of habitats as possible and should incorporate some reassessment of the taxonomy of the organism.

5. Experiments in adding and subtracting specific organisms or groups of organisms (e.g., all ants, all insects, all vines) from many of the types of communities are needed. This should be done both to search for and to test hypotheses.

6. Taxonomic treatments focused on prominent groups in badly disturbed habitats should be encouraged. These should proceed hand-in-hand with the development of easily accessible data-retrieval systems for all the information available concerning particular organisms.

7. Special attention should be given, in the above studies, to (a) the evolutionary and contemporary origin of the organisms found in changing habitats, (b) the resources of energy, manpower, and materials required to initiate, accelerate, or delay successional changes, (c) the probability of various types of outcomes from seemingly similar types of disturbance, (d) the effects that successional changes have on organisms and communities outside the one directly concerned.

8. Knowledge of soil changes occurring in the course of secondary succession is needed. Studies should be made of changes in total storage of fertility elements such as total carbon, nitrogen, exchangeable cations (calcium, magnesium, potassium, sodium), exchange capacity and phosphorus, as well as various trace elements. There are many soil data available in the tropics that have not been adequately integrated into the ecological framework of natural vegetation and successional change studies. These data should be utilized in ecological studies to save needless repetition of expensive laboratory work. Special studies are needed of soil factors that guide succession, e.g., studies of mineral element deficiencies, interactions with soil pathogens, and the role of mycorrhizae in tropical forests.

9. Detailed publication of data from the studies mentioned above, in journals or other publications of wide circulation (not only data banks), is needed. It is imperative that these receive widespread translation into Spanish, English, and Portuguese.

III. Existing Knowledge and Lacunae

Of the many ways to approach research in the field of secondary succession, two are basic. One is to work in a single area with similar climate and soils and study a series of vegetation stands at different ages of recovery, preferably a whole series including a sample of the original community. This is the conventional method Budowski (1961), Gómez-Pompa (1966), and Kellman (1970a) employed in tropical regions, and it has produced useful results. Unfortunately, the sampling procedure, the uncertainty of the site history (Kellman, 1970a), the data gathered, and the reliability of the species identifications have often been so unsatisfactory that regional comparisons cannot be made.

To our knowledge the only extensive tropical survey of this kind with adequate data for comparisons is the one made in the humid lowlands of the Gulf of Mexico (Instituto Nacional de Investigaciones Forestales, 1970-1971) which is still being interpreted. Secondary successional studies in Mexico began with the work of a group of investigators headed by Faustino Miranda, and from it emerged important early papers by Miranda et al. (1960a,b). This research was the first in Mexico to use direct observations in an attempt to integrate different successional stages in a tropical lowland. Sousa (1964) was concerned with identifying "indicator species" that could enable an ecologist to determine the age of a secondary stand as well as the nature of the original vegetation prior to the disturbance. Computer analyses of structural variation during secondary succession also have been attempted (Sarukhán and Gómez-Pompa, 1963; Rico et al., 1972). Studies of successional sites of known origin have been carried out by Sarukhán (1964).

Present interest in the regeneration of tropical rain forest in Mexico has been directed toward rates of speciation in secondary communities (Gómez-Pompa, 1971b), influence of seeds in the soil of the primary forest on the course of succession (Guevara and Gómez-Pompa, 1972), the ecophysiology of seed germination of selected secondary species (Vázquez and Gómez-Pompa, 1971; Vázquez, 1973), structural and compositional changes during the early stages of recovery of experimentally disturbed plots, and the use of computers in the processing of the copious data provided by such studies (Rico, 1972).

A second approach to the study of secondary succession is to control the origin of the disturbance by experimental methods. An area is disturbed in a certain way and the succession is studied through lapses of time. This approach is undoubtedly the most precise but it has (up to now) great limitations as the process requires many years, during which researchers (or administrators) may change their interests and terminate the project. Many weaknesses mentioned for the first approach apply to this one as well, with the additional disadvantage that this second approach has been utilized in only a few places. It is clear that research of this type should be coordinated at higher levels of administration to insure continuity and to make sure that the research strategy, the experimental design, and the data collected are adequate for comparisons and for advancing our knowledge of the process.

With these two basic approaches in mind, what are the main steps to be taken and what types of data should be gathered to gain an understanding of secondary succession in the tropical lowlands?

A. Major Factors Initiating Succession

Succession is initiated by disturbance that substantially destroys the structure of the primary community. These disturbances include fire, shifting agriculture, flooding, landslides, and herbicidal defoliation. Although primary communities are the basic starting point for the study of secondary succession, the actual communities observed in the field are often derived from other second-growth stands. In fact, it is common practice in shifting agriculture to allow a fallow period of second growth to intervene between the cycles of crops.

It is clear from numerous studies of succession, where fields with different histories but common ages are compared, that past use is a strong determinant of the successional process. For example,

Kellman (1969) found in Mindanao that the above-ground vegetation mass varied as much as five times in plots of the same age (Table 4.1) but with different histories. The number of species were also quite different (on 7-year-old fields the number of species varied from 20 to 76).

TABLE 4.1. Above-ground biomass of second-growth vegetation in Mindanao, Philippines

Age of Field	Fresh Biomass (kg ha^{-1}x 10^{-3})
1.0	31
1.0	23
2.5	39
3.0	27
6.5	119
7.0	117
7.0	83
7.0	27
19.0	136
19.0	116
19.0	103
19.0	81
21.0	134
23.0	84
27.0	419
27.0	370
mature	11,205

Data from Kellman (1970a).

It appears from the few studies available that the size of area disturbed and the frequency of repeated disturbance are major determinants of the succession process. On very large areas, the invasion of disseminules must necessarily take place from the edge of the plot so that full stockings of the stand may be greatly delayed. Repeated disturbance results in destruction of disseminules present in the soil, as well as the reduction of the ability of the soil to support plant growth. Repeated disturbance commonly causes a deflection from the primary community endpoint toward another community type. Thus, it is common in tropical areas for formerly forested areas to be converted into very long-lived grass or savanna communities through repeated disturbance.

Fire is among the most important of the physical factors influencing succession, and, although we have a large body of descriptive information on the effects of fire in certain communities (Batchelder and Hirt, 1966; Bartlett, 1955, 1957, 1961), we lack information about its effects on the successional process. Factors such as intensity of fire, temperature, frequency of burning, soil temperature during the fire, and effects of high temperature on propagules and on physical and biotic properties of the soil can be critical in determining the direction of the succession.

The existence of recurrent fires over large areas has been used to explain the development of herbaceous pyrophilous communities, which interrupt the succession normally leading to a forest stand, (Budowski, 1961; Daubenmire, 1972). Information is lacking on succession on pyrophilous communities in the American tropics following the cessation of annual fires.

The impact of various disturbances, especially in the early stages of succession, is easily amenable to experimental treatment (Odum, 1969) and should be studied in each major community since we suspect that typical dry-climate communities may be very different than moist climates in their capacity to recover. In these experimental studies it is desirable that a comparison be made of the successional stage with the primary community. However, this requirement demands adequate protected stands for study.

The following general features should be represented within these areas: (a) They should be of adequate size and pattern to maintain the primary community in the face of depauperization due to insular effects (see comments in Section 2); (b) they should be of adequate size to encompass the full range of natural perturbations, such as hurricanes and the effects of primitive man; (c) They should contain a number of watersheds to facilitate studies on the effects of secondary communities on hydrologic and geomorphologic landscape characteristics; (d) They should be of adequate size to allow experimental perturbations within them without extensive effects upon the preserved primary communities.

B. A Description of Successional Processes

An explanation of succession should describe the changes in community structure over time, explain how these changes occur, and determine if the steady state resulting from the successional process is unique or is one of a set of states characteristic of the region. The fundamental operational problem in the study of succession is the description of the rate of change in community structure and function. Not only are the descriptive data essential for construction of experimental hypotheses but the distinction between successional and steady-state communities depends upon the interpretation of the rate processes, since successional communities change rapidly while steady-state communities change slowly (at least, over the evolutionary and geological time span).

Whatever the structural or functional parameters chosen for study, the typical model describing these processes is an ascending curve following disturbance (Fig. 4.1). The initial condition is perturbed, the process or state is depressed dependent upon the disturbance, and once the disturbance is removed the process begins to change toward an endpoint as mentioned earlier. The history of the plot or the nature of the disturbance may result in the community reaching different endpoints. For example, the succession in a forest opening or a small field cut from the original forest and burned only once will probably move directly to the primary community type. In another case, repeated disturbance of the site results in a succession that is deflected or, in the worst case, arrested at another endpoint.

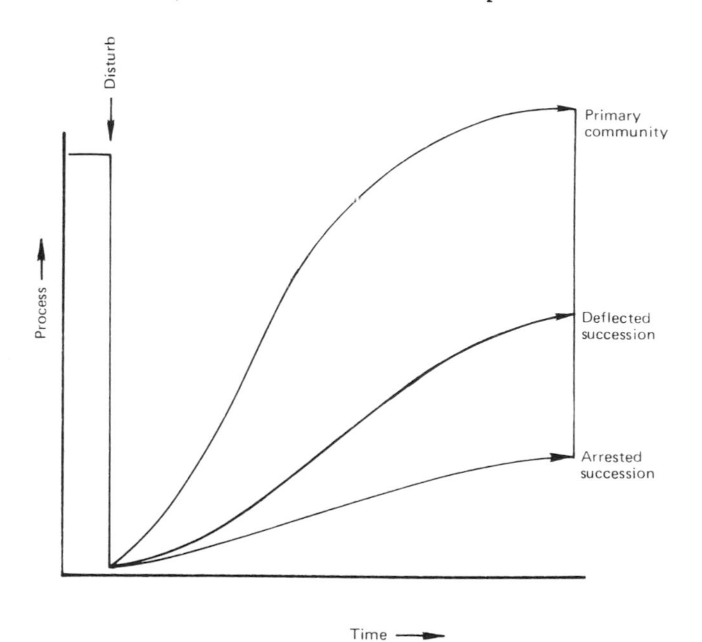

FIG. 4.1. Model of succession processes. This diagram describes only one of many processes in successional systems that could reach an endpoint within the limits described by the upper and lower lines.

The development of the vegetation early in the sere is influenced by the presence or absence of root stocks and stumps, buried seeds, and dissemination of seeds from outside the plot by wind and animals.

The first two factors are affected by the cultural history of the plot and the latter by the distance to seed sources. The subsequent development of the community is a function of the environmental factors and the nature of the biota.

Structural changes in the community include alterations in numerous parameters, including its composition, variation in height and life form, biomass, energy and mineral content, and soil structure. Richards (1952) described tropical secondary forest vegetation as lower in height and smaller in diameter than primary forest, as well as being more uniform and regular and having fewer species. He stated that the dominant plant species in tropical terrestrial succession are often rapid-growing and shade-intolerant (e.g., members of *Cecropia*, *Vismia, Ochroma,* and *Trema*). Budowski (1965) has developed an organized listing of these characteristics of succession for land managers.

Studies of succession, mainly in temperate regions, have shown that following dominance by a few species early in the sere there is an overall increase in species numbers with time. Numerous studies of the tropical successional vegetation (Ross, 1954; Lawson et al., 1970; Kenoyer, 1929; Kellman, 1970a; Symington, 1933; Sarukhán, 1964) suggest that while the overall species richness is greater in tropical successional communities, the rate of increase in species richness is not basically different from temperate communities. This conjecture should be examined further by comparative studies in both regions—apparently it is due to a greater number of species present at the initiation of the sere in the tropical environment.

The weights of tropical secondary vegetation have been summarized by Snedaker (1970) and others. Biomass increases with time and eventually reaches as much as 1000 mt ha^{-1}. However, large variation between plots at a given age (Table 4.2), especially at early stages of succession (Kellman, 1970a; Snedaker, 1970), reflects the effect of environmental factors, history of the site, and the biota.

The change in biomass is reflected in changes in chemical energy or elemental content of the community and soil, although there are few data on these topics. Concentrations of macronutrients vary widely between sites (Tergas, 1965; Snedaker and Gamble, 1969; Stark, 1971; and Golley et al., in press). For example, in a primary forest in Panama the pattern of concentration of the more abundant elements (excluding C, H, O, N) are Ca > K > Mg > P > Al. In second growth (2 to 6 years), at the same site the pattern is K > Ca > Mg > P > Al. The reverse of calcium and potassium apparently is due to greater quantities of calcium in cell walls in the mature forest. Many more comparative data on energy and mineral content are needed to determine the nature of these differences between primary and secondary forests.

Functional processes in tropical succession include productivity, mineral cycling, community stability, and phenology. In general, these

TABLE 4.2. Variation in biomass of successional vegetation at Lake Izabel, Guatemala

Age	N	Mean	Range
10 mo	10	9,415	4,430- 14,140
1 yr	17	8,327	3,870- 15,690
2	16	14,199	5,600- 27,560
3	7	22,820	9,480- 46,240
4	16	26,908	12,920- 48,180
5	16	36,666	12,210- 84,420
6	16	44,669	14,730- 70,160
7	16	46,658	14,400- 96,830
8	4	65,823	29,070- 97,930
9	4	72,403	26,090-114,240
10	32	53,303	10,000-210,320

Data from Snedaker (1970).

functions are less understood than structural changes. If we can extrapolate from the generalizations of Kira and Shidei (1967), based mainly on temperate studies but including tropical data, primary production increases with time, reaching a peak before the mature community is achieved. The biomass values for tropical second growth also suggest an initial rapid period of production and then a decline in production rate, although no sere has been adequately studied from this point of view. We require many more observations on production of successional communities, since temperature, moisture, solar energy, and soil fertility all directly influence the process (Lieth in Golley and Lieth, 1972).

The rate of mineral cycling has been determined for very few tropical systems of any age, although we do have a number of studies of leaf fall, which give a crude estimate of mineral cycling since they measure a major part of the annual loss from the vegetation to the soil. Leaf mass is re-established early in the sere (Table 4.3) (Golley et al., in press), and, therefore, the leaf litter fall should reach a steady level early in succession. However, mineral uptake of the vegetation should continue to increase since the wood is increasing in amount. The mineral turnover of the more active part of the community (leaves, fruits, and animals) can be considered separate from the less active woody portion, which reaches equilibrium at a much later time.

The minerals comprising the successional community are derived mainly from the soil, although atmospheric or hydrographic inputs may be locally important. Where the soil nutrient capacity is low, redevelopment of the community may be retarded. Stark (1971) suggests that this is the case for podzolic sands in the Amazon basin. However, there is also a reciprocal effect of the community on the soil (Kellman, 1969; Popenoe, 1957; Zinke et al., 1970, and others). Soil carbon increases rapidly with succession, whereas soil nitrogen increases more slowly. Kellman (1969) suggests that successional species cycle elements in greater amounts than primary species and that specific plants (*Trema orientalis* and *Melastoma polyanthum*) restore phosphorus and potassium to the upper soil layers. Clearly, the mineral cycling and availability may be central to the successional process and should be studied intensively.

TABLE 4.3 Leaf area, leaf mass, and total biomass versus age of stand in a tropical moist successional forest in Panama

	Age of stand (yr)			
	2	4	6	Mature
Leaf area index	8	12	17	11 - 22
Leaf mass (mt ha^{-1})	4	6	7	8 - 12
Total biomass	16	42	57	276 - 378
Ratio of leaf mass to total biomass	0.25	0.15	0.12	0.05

Data from Golley et al. (in press).

Odum (1969) suggests that succession results in an increase in community stability (i.e., the ability of the community to resist disturbance). There are no studies to test this hypothesis in tropical communities but it is subject to direct experimentation. Herbicidal treatment, for example, may be a tool to test the ability of the community to resist disturbance (Ewel, 1971b).

Finally, the timing of biological processes, including reproduction and growth, may change with succession in tropical communities. The proximate environmental cues that affect organisms often are moisture or light changes. We assume that the greater species variety in mature communities would correlate with increased heterogeneity in response to these cues as they are used by species in the allocation of resources. We would expect successional communities to have more regular timing (Richards, 1952).

Succession is a community phenomenon in which the community structural and functional features change over time so that it appears that there is a regular, and possibly a predictable, series of events leading ultimately to a steady-state community, or at least to a community that changes at an imperceptible rate. The overall description of structural and functional dynamics in tropical successional communities is poorly known and basic ecological research on these topics is well worth attention. However, description of the successional processes is not sufficient to understand succession and manipulate it. The structural and functional changes result from interactions between the components of the community—the species populations, trophic groups, and individuals—and these must be understood to explain succession.

C. Interaction of Successional Components

There are several major types of interactions that were identified in Section 2 and that are also important in the study of succession. Among these are interactions of the biota with environment, plant-plant interactions, and plant-animal interactions.

1. Effect of the environment

It is of great interest to compare succession in different tropical climates since the physical environment will have a strong influence in determining which species populations can establish and grow. Succession in tropical areas with highly seasonal rainfall is expected to be quite different from that in regions with more evenly distributed rainfall, since plants in the early stages will have to survive several months of water shortage. Altitudinal and latitudinal gradients in tropical ecosystems are also climatically controlled, but almost nothing is known about the way the gradients affect the successions. At low and medium elevations (500 to 2500 m), as in the lowlands, large areas of forest and other natural communities have been destroyed or damaged by man or replaced by cultivation and grazing land. However, there is little information on the successions taking place on abandoned cultivated land and few of the secondary communities have ever been described. Tropical succession is extremely slow at high elevations (above 2500 m). When páramo or subpáramo vegetation is cleared by fire, grazing, or other catastrophic events, recovery takes a long time. In the early stages of this succession, the ground may remain completely free of higher plants for years. We suspect that this is due to very slow growth rate of seedlings rather than a lack of seed input. Even more striking is the extremely slow rate of decomposition of plant remains following disturbance. It is interesting to note that there is also total absence of large decomposer insects (wood-boring beetle, ants, termites) in these areas.

2. Influence of the biota on succession

There is a reciprocal effect of the biota on the physical environment in the successional community, although we have very little information on this feedback. The amelioration of changes in temperature, moisture, height at the soil surface and in the soil undoubtedly have a strong influence on the germination of seeds and the growth of plants. These microenvironmental impacts are probably most important in the early stages of succession, when the community has not developed sufficiently to buffer adequately the biota from the direct impact of the physical environment.

Scattered data on secondary vegetation in the tropics suggest that the composition and origin of the flora and fauna available to disturbed sites is important in determining the course of succession. Major influences on the biota appear to be the degree of disturbance to the site prior to abandonment and its location relative to other communities. A fuller understanding of these biotic factors is necessary, and will require not only more detailed observations but also an experimental approach.

The organisms that invade a disturbed habitat come from two sources: the remnant populations of individuals of many ages that have survived the disturbance and immigrants that arrive at the site after the disturbance. Once established, the community may become a major contributor of propagules to later phases of the succession, although these may not survive until the disturbance is repeated. Although the migrant source of propagules traditionally has been assumed to be the major input of organisms in succession, clearly the remnants of previous populations may be of even greater importance in areas that are repeatedly disturbed. For example, stumps and root stocks capable of regenerating a standing plant population and a resident population of viable seeds in the soil may produce the largest proportion of living plants in the sere. The importance of stumps and surviving roots in the regeneration of fallow in shifting cultivation has often been given passing mention in the literature (e.g., Emerson, 1953; Sarukhán, 1964; Kellman and Adams, 1970), but no quantitative studies appear to have been carried out on this interesting phenomenon. There may be wide regional variations in importance of root stocks even under comparable types of disturbance.

Buried viable seeds are a second potential component of the remnant populations. Most research on the phenomenon has been undertaken in temperate areas, where a large population of buried seed of weedy and other secondary species exists. Comparable data from the tropics are scarce. Only Symington (1933) in Malaya, Keay (1960) in Nigeria, Bell (1970) in Puerto Rico, and Guevara and Gómez-Pompa (1972) in Mexico have treated the phenomenon directly. Their results suggest that a large population of buried viable seed of successional species may exist in tropical soil (Table 4.4). Circumstantial evidence for the existence of seeds in the soil also comes from the fact that repeated weedings appear to deplete the secondary tree population (Kellman, 1970b; Havel, 1960). The ability of seed of some tropical secondary species to retain viability in the soil for at least 6 years has been demonstrated by Juliano (1940).

The significance of remnant populations to succession lies in their potential for rapid establishment at a site early in the succession. If residents are absent, a site may become occupied by an entirely migrant community, effectively suppressing subsequent vegetation changes for a prolonged period.

Factors controlling the size and composition of remnant populations are beginning to be understood. At the minimum, fire, weeding frequency, grazing, and ploughing all appear important in the determination of these populations (Brinkmann and Vieira, 1971; Havel, 1960; Kellman, 1970b; Kellman and Adams, 1970; Wyatt-Smith, 1949). On the other hand, the origin of the buried viable seed population and its longevity is unknown.

3. Plant-plant interactions

Temporal changes in species mixtures in successional communities are commonly the result of competitive interactions between populations and individuals. Plant-plant competition comprises a large proportion of such interactions. The generally observed inability of pioneer species to regenerate in their own (or another species') shade is an obvious example. Beyond generalizations at this level we have practically no information about how interactions between plants, either competitive or mutualistic, affect the course of succession in the tropics. Experimental work on temperate crop plants has demonstrated the subtlety of plant-plant interactions. It is not enough to identify an interaction for we need information on the morphological, physiological, and reproductive behaviors of the interacting plants. We may expect to find that timing and intensity of reproductive efforts, growth rate of the whole plant, partitioning of this growth rate between the growth of different functional parts of the plant body and chemical substances released by the plants are important (Anaya, 1973).

4. Animal-plant interactions

Animals alter competitive outcomes between plants within a successional community. This may change species composition, by seed predation (Janzen, 1970) and the rates of change and transitional probabilities from one stage to another. These effects are mediated through seed predation, herbivory, and sometimes by trampling.

Animals (primarily vertebrates) transport seeds within a successional community and between successional communities of different

TABLE 4.4 Summary of available data on buried viable seed populations in tropical soils

Area	Vegetation Type	Sample Area (cm^2)	Sample Depth (cm)	Burial Viable Seed Population	Reference
Malaya	"Edge of the jungle"	2 x 40.481.44	2.5 ("layer of humus.")	'Masses of composite weeds, grass and creepers. . . . Musa. Trema, Macaranga, etc."	Symington (1933)
Veracruz	Secondary vegetation 2 mo old	16 x 39.75 (x 8 repeats)	12	23 spp.; 862-2672 seedlings/m^2	Guevara and Gómez-Pompa (1972)
	Secondary vegetation 5 yr old	16 x 39.75 (x 8 repeats)	12	19 spp.; 1982-3879 seedlings/m^2	
	Primary forest 5 yr old	16 x 39.75 (x 8 repeats)	12	26 spp.; 344-862 seedlings/m^2	
	Primary forest	16 x 39.75 (x 8 repeats)	12	13 spp.; 175-689 seedlings/m^2	
Puerto Rico	Primary montane rain forest	6 x 2500	7-10	15 spp.; 152-424 seedlings/m^2	Bell (1970)
Belize	Pasture and cultivated fields	78 x 29.2	4.2	54 spp.; 6497 seedlings/m^2 (mean)	Kellman (unpblished)

ages. Some of these animals are seed predators as well as dispersal agents (Janzen, 1971). However, it is unlikely that a direct and simple relationship exists between the numbers and kinds of seeds carried by animals and the consequences of this movement to the successional community as a whole or to particular species within it. Seed predation by animals is a major cause of the loss of seeds from plants, litter, and soil. However, we can only infer this from the fact that many animals found in secondary communities (peccaries, rodents, ants, beetles, bugs) subsist in great part on seeds.

Animals, particularly insects, carry pathogens among plants. This may be of particular importance in secondary communities where plants often form extensive stands of low species diversity. Herbivory by animals could weaken plants and make them more susceptible to diseases. Again, this may be more noticeable in depauperate successional communities when insect outbreaks occur.

Animals also influence the rates of vegetational change by altering the rates of litter decomposition and litter fall. Over evolutionary time, animals may also limit the possible plant strategies, because their presence requires certain minimal expenditures by plants for chemical or other defenses.

Animals parasitize and prey on other animals, influencing succession in the same way that man does when he experimentally alters successional animal communities.

From the above list it is clear that animals, in general, may have the same types of effects on secondary succession as they have in less disturbed habitats. However, the vegetation of secondary successional habitats differs in a number of ways from that of undisturbed vegetation. Whereas many of these characteristics have been mentioned earlier (Richards, 1952; Budowski, 1961), the fact that successional communities may be disturbed repeatedly over large areas and that the various disturbances, inputs, and organisms associated with the vegetation deviate from the usual patterns all may have an impact on animal-plant interactions. This is an area that merits much detailed research since it has an obvious importance to biological control and maintenance of cultivated crops.

5. Role of autecology in understanding succession

A voluminous literature pertaining to the ecology of successional species exists, but most of this information consists of natural-history observations or results of intensive investigations of a rather narrow scope. Most of these studies are on economically valuable species, such as timber trees or agricultural crops and weeds. Even though many economically valuable species occupy a successional position under natural conditions, the accumulation of this information has not appreciably aided our understanding of succession. We believe that this situation exists for two reasons: (a) the only common ground is the species, resulting in a potpourri of unrelated and often unsynthesized information drawn from many habitats, and (b) most such studies lack a contextual frame of reference, i.e., we know extremely little about the status of species in succession. This is not to imply the irrelevance of "natural-history" type information. On the contrary, it can be very useful in generating hypotheses and providing basic ecological information (Gómez-Pompa, 1971a). Indeed, a well integrated autecological study is often referred to as an ecological life history. Although very few such studies have been conducted in the American tropics (Hartshorn, 1972; Gliessman, 1973), ecological life history studies could greatly improve our understanding of succession. An especially good example is the "Biological Flora of the British Isles" series published in the Journal of Ecology. Such intensive studies of temperate species should serve as a guideline for the development of studies of tropical successional species.

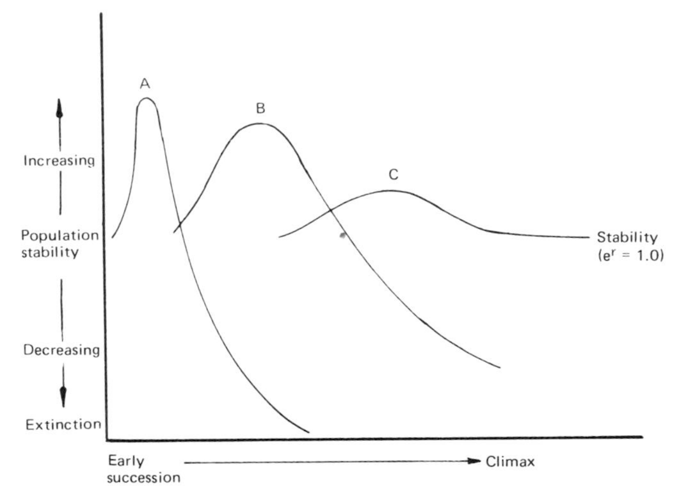

FIG. 4.2. Hypothetical rates of increase for (A) early successional species, (B) late successional species, and (C) climax species on a successional gradient.

Two particular aspects of autecological study have been virtually ignored: population dynamics and population-nutrient relationships. The difficulty (see Budowski, 1965) of distinguishing between late-successional and climax communities could be greatly reduced, if not resolved, with population-dynamics studies of important or indicator species. A mathematical model (Hartshorn, 1972, 1974) could be used to determine the stability of numbers and size-class distribution of each population. If the populations studied are stable, then a strong argument could be made for the equilibrium status of the community.

In contrast, instability of population numbers and/or size-class distribution may indicate an unstable or successional community. With this type of approach, it might also be possible to arrange a group of communities or forest stands along a successional gradient based on the intrinsic rate of increase of indicator species (Fig. 4.2). For example, if we know that species A is an early successional species under certain edaphic and climatic conditions and we obtain different measures of the rate of increase of a population for several stands containing species A, we can ordinate these stands along a successional gradient, where the value of the rate of increase of A decreases as succession progresses.

Autecological studies of the relationship of early successional species to nutrient cycling could be a particularly productive area in the quest for understanding succession. The following questions are examples of the types of problems that will prove interesting. Are the very fast-growing weedy species, such as *Ochroma, Trema, Cecropia, Musanga,* more effective at capitalizing on a large but temporary nutrient resource than are agricultural species? Do these species store nutrients that would otherwise be lost from the system?

6. Edaphic aspects of secondary succession

The edaphic aspects of secondary succession in the tropics is of special interest since soil fertility and soil structure have a controlling influence on plant growth. Soil serves as the substrate for vegetation growth. It is the supply source of essential elements and the medium for reincorporation of chemical elements upon death of the vegetation. Finally, soil stores essential elements following weathering of base rock. Soil also is the habitat of a microflora and microfauna that interact with the macrovegetation of the surface. In all of these aspects soil can affect the processes of secondary succession.

Vasconcelos (1971) categorized the edaphic factors affecting the sequence of vegetation on a site as the nutrient-element supply, the soil drainage and moisture relationships, the soil temperature, and the soil microflora and fauna. Schnell (1971) reviewed the control of soils on tropical vegetational processes. He points out that local soil conditions may play a primary role in the mosaic of differences in vegetation (Fig. 4.3). An example of this influence is the distribution of laterite crusts, where the successional sequence will be very different from that on adjacent latosol areas. In Cambodia, Schnell describes fire-maintained savanna grasslands correlated with the xeric soil moisture regimes on shallow laterite crusts. Near Santarem, in the Amazon Valley, a lateritic concretion layer in the soil influences the dynamics of the forest and the successional sequence that takes place after harvest. This is partially related to the physical characteristics of the soil and attendant water-storage capacities, and partly to nutrient problems such as phosphorus fixation in the soil. The soil

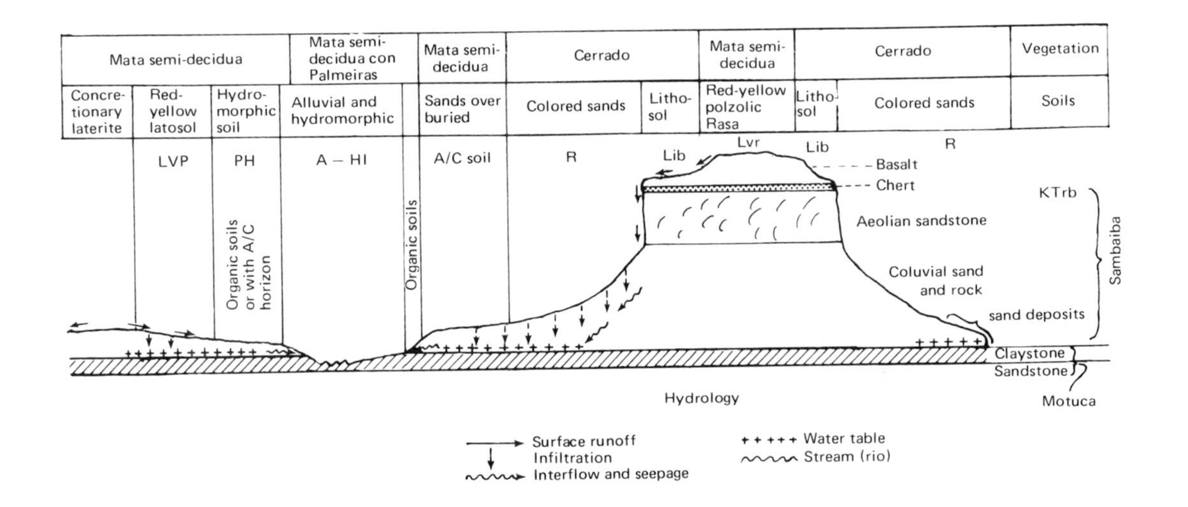

FIG. 4.3 Successional sequences vary depending upon local enviornmental variables. This area near Carolina in the state of Maranhao, Brasil illustrates a local complex of soil, vegetation, geology, geomorphology, and hydrology. These features result in a mosaic of successional sequences following disturbance on each type. (From Zinke 1971.)

fauna also play a role in the interaction between soil and the secondary succession. For example, Pendleton (1941) described various species of plants found on termite mounds relative to those on adjacent areas in abandoned fields.

The control of soil types on secondary succession was well documented in older settled areas in the tropics by DeCamargo (1948). Detailed descriptions of species successions on various soils following cutover in the so-called Capeoira forest in Brazil are numerous in the soils literature, an example being that by Vieira et al. (1967). In such areas as the white sands, the sequence of secondary succession is very rigorously controlled by the soils (Klinge, 1968). The relation of soils to the geologic substratum has been shown by Veloso (1972) in vegetation analysis of quadrangles of the Radar Map of Amazon Vegetation. For example, he described secondary succession to a vine- and creeper-dominated forest (Cipoal) as tending to occur on soils derived from Ipuxuna and Precambrian formations, whereas *Dinizia* and other rainforest species are maintained on adjacent tertiary formations without Cipoal. These differences are correlated with differences in soil texture, nutrient supplying power, and drainage, all related to the geologic substratum. Further discussion of soils under the shifting cultivation cycle can be found in Section 5, under agricultural systems.

The ultimate effect of a disturbance of vegetation on secondary succession involving soil will be in the process of actual erosion and disappearance of the soil and its nutrient base. As Birot (1968) emphasized, the basic erosional cycle of tropical climates is the production of rounded land forms with chemical weathering of the rock and soil. The result is very fine-grained sediment with deeply weathered surfaces. Severe disturbances on steep slopes may interrupt this process resulting in coarse sediments and shallow soils. Entirely different successional sequences will result on these altered surfaces.

Finally, the soil may enter chemical phases as a result of disturbance that produces altered successions. For example, a mangrove-forest soil may be turned into an acid sulphate soil with a very low pH resulting in a secondary succession to a *Melaleuca*-reed grass savanna, or even bare ground. Removing a forest may in certain circumstances enhance hardening of laterite.

These examples of interactions between soil and succession in the tropics do not represent a complete review. There are many other details of specific interest such as the role of mycorrhizae in tree nutrition on nutrient-poor tropical soils. Many specific deficiencies of trace elements on tropical soils occur on older surfaces and probably have a role in the selection of species occurring in successional sequences. Also, there are allelopathic interactions between the remains and excretions from vegetation growing on the soil inhibiting secondary succession species (Anaya, 1973). The role of species with special functions such as nitrogen fixation or those with enhanced phosphorus uptake from soils having fixed phosphorus or the special cases of waterlogged soils with their anaerobic environments, and the effects of extremes in the soil moisture regime from drought to waterlogging are further examples of soil-succession interactions. These interactions between soil and the vegetational changes that will occur illustrate the varied nature of soil interactions and the continuous need to be aware of their possibilities in the study of secondary succession in the tropics. Research on succession must include the edaphic factor.

References

Anaya, A. 1973. Algunos aspectos importantes de la ecologia quimica en las regiones tropicales, unpublished manuscript.

Bartlett, H. H. 1955. Fire in relation to primitive agriculture and grazing in the tropics. Annotated Bibliography. University of Michigan, Ann Arbor. Vol. I, 568 p.

Bartlett, H. H. 1957. Fire in relation to primitive agriculture and grazing in the tropics. Annotated Bibliography. University of Michigan, Ann Arbor. Vol. II, 873 p.

Bartlett, H. H. 1961. Fire in relation to primitive agriculture and grazing in the tropics. Annotated Bibliography. University of Michigan, Ann Arbor. Vol. III, 216 p.

Batchelder, R. B., and H. F. Hirt. 1966. Fire in tropical forests and grassland. U.S. Army Natick Lab. Tech. Rep. 67-41, ES. 380 p.

Bell, C. R. 1970. Seed distribution and germination experiment, p. D-177-D-182. *In* H. T. Odum and R. F. Pigeon (eds.) A tropical rain forest. A study of irradiation and ecology at El Verde, Puerto Rico. U.S. Atomic Energy Commission, Oak Ridge, Tennessee.

Birot, P. 1968. The cycle of erosion in different climates. University of California Press, Berkeley. 144 p.

Brinkmann, W. L. F., and A. N. Vieira. 1971. The effect of burning on germination of seeds at different soil depths of various tropical tree species. Turrialba 21: 77-82.

Budowski, G. 1961. Studies on forest succession in Costa Rica and Panama. Unpublished Ph.D. Thesis, Yale University, New Haven, Conn. 189 p.

Budowski, G. 1963. Forest succession in tropical lowlands. Turrialba 13: 42-44.

Budowski, G. 1965. Distribution of tropical America rain forest species in the light of successional process. Turrialba 15: 40-42.

Daubenmire, R. 1972. Some ecological consequences of converting forest to savanna in Northwestern Costa Rica. Trop. Ecol. 13: 31-51.

De Camargo, F. C. 1948. Land and settlement on the recent and ancient quaternary along the railway line of Braganca, State of Para, Brasil, p. 213-221. *In* Proc. Inter-Amer. Conf. Conserv. Renewable Natur. Resources. Dept. of State, Washington, D.C.

Emerson, R. A. 1953. A preliminary survey of the milpa system of maize culture as practiced by the Maya Indians of the northern part of the Yucatan peninsula. Ann. Missouri Bot. Garden 40: 51-62.

Ewel, J. 1971a. Cambios de la biomasa en la sucesión temprana de bosques tropicales. Turrialba 21: 110-112.

Ewel, J. 1971b. Experiments in arresting succession with cutting and herbicides in four tropical environments. Unpublished Ph.D. Thesis, University of North Carolina, Chapel Hill.

Foggie, A. 1960. Natural regeneration in the humid tropical forest, p. 1941-1946. *In* Proc. Fifth World Forestry Congress. University of Washington, Seattle.

Gliessman, S. R. 1973. The phytotoxic potential of bracken in relation to land management problems in the humid tropics. Unpublished manuscript.

Golley, F. B., and H. Lieth. 1972. Bases of organic production in the tropics, p. 1 to 26. *In* P. M. Golley and F. B. Golley (compilers) Tropical ecology with an emphasis on organic production. University of Georgia, Athens.

Golley, F. B., J. T. McGinnis, R. G. Clements, G. I. Child, and M. J. Duever. 1974. Mineral cycling in a tropical moist forest ecosystem. University of Georgia Press, Athens, in press.

Gómez-Pompa, A. 1966. Estudios botánicos en la región de Misantla, Veracruz. Ed. Inst. Mexicano de Recursos Naturales Renovables, A. C. México. 173 p.

Gómez-Pompa, A. 1971a. Posible papel de la vegetación secundaria en la evolución de la flora tropical. Biotropica 3: 125-135.

Gómez-Pompa, A. 1971b. Ecología de una especie tropical, p. 105-108. *In* A. Gómez-Pompa y S. del Amo R. (eds.) Problemas de investigación en botanica. Limusa Wiley, México.

Gómez-Pompa, A., and L. Cazares. 1970. Mapas de vegetación en zonas cálidas y su importancia. Bol. Esp. Inst. Nac. Invest. Forest. México 5: 1-11.

Gómez-Pompa, A., M. Sousa S., and J. Sarukhán K. 1963. La vegetación secundaria en las zonas cálidas mexicanas, p. 19. *In* Resúmenes de los trabajos. II Congreso Mexicano de Botanica, St. Luis Potosi, Mexico.

Gómez-Pompa, A., J. Vázquez Soto, and J. Sarukhán K. 1964. Estudios ecológicos en las zonas tropicales calido húmedas. Publ. Esp. Inst. Nac. Invest. Forest México 3: 1-36.

Gómez-Pompa, A., C. Vázquez-Yanes, and S. Guevara. 1972. The tropical rain forest: A nonrenewable resource. Science 177: 762-765.

Guevara, S., and A. Gómez-Pompa. 1972. Seeds from surface soils in a tropical region of Veracruz, Mexico. J. Arnold Arboretum 53: 312-335.

Hartshorn, G. S. 1972. Ecological life history and population dynamics of *Pentaclethra macroba* a tropical wet forest dominant and *Stryphuodendron excelsum,* an occasional associate. Unpublished Ph.D. Thesis, University of Washington, Seattle.

Hartshorn, G. S. 1973. A matrix model of plant population dynamics. In F. Golley and E. Medina (eds.) Tropical Ecological Systems. Springer-Verlag, New York, in press.

Havel, J. H. 1960. The deflection of secondary succession on cleared mid-mountain rainforest sites by plantation tendings, p. 339-343. *In* Symp. on the Impact of Man on Humid Tropics Vegetation, Goroka TPNG, Sept. 1960, Canberra, Australia. Commonwealth Government Printing Office, Canberra, Australia.

Instituto Nacional de Investigaciones Forestales (INIF). 1970-1971. Comision de estudios sobre la ecologia de Dioscoreas. Informes 1-6, Pub. de tiraje restringido del Instituto Nacional de Investigaciones Forestales., S.A.G., Mexico.

Janzen, D. H. 1970. Herbivores and the number of tree species in tropical forests. Amer. Natur. 104: 501-528.

Janzen, D. H. 1971. Seed predation by animals. Ann. Rev. Ecol. Syst. 2: 465-492.

Juliano, J. B. 1940. Viability of some Philippine weed seeds. Philippine Agr. 24: 313-326.

Keay, R. W. J. 1960. Seeds in forest soil. Nigerian Forest. Infor. Bull. NS. 4.

Kellman, M. C. 1969. Some environmental components of shifting cultivation in upland Mindanao. J. Trop. Geogr. 28: 40-56.

Kellman, M. C. 1970a. Secondary plant succession in tropical montane Mindanao. Dept. Biogeogr. Geomorph. Publ. BG/2 Australian National University, Canberra, Australia.

Kellman, M. C. 1970b. The viable seed content of some forest soil in coastal British Columbia. Can. J. Bot. 48: 1383-1385.

Kellman, M. C., and C. D. Adams. 1970. Milpa weeds of Cayo District, Belize (British Honduras). Can. Geogr. 14: 323-343.

Kenoyer, L. A. 1929. General and successional ecology of the lower tropical rain-forest of Barro Colorado Island, Panama. Ecology 10: 201-222.

Kira, T., and T. Shidei. 1967. Primary production and turnover of organic matter in different forest ecosystems of the western Pacific. Jap. J. Ecol. 17: 70-87.

Klinge, H. 1968. Report on tropical podzols. FAO Report. 88 p.

Lawson, G. W., K. O. Armstrong-Mensah, and J. B. Hall. 1970. A catena in a tropical moist semi-deciduous forest near Kadi, Ghana. J. Ecol. 58: 371-398.

Miranda, F., A. Gómez-Pompa, L. Henandex, and M. Rodriguez. 1960a. Estudio de la vegetacion sobre cerros carsticos en Tuxtempu, Oax, p. 68. *In* Resumenes de los trabajos prenentados. I Congreso Mexicano de Botanica, Mexico.

Miranda, F., S. Vázquez, J. Gonzalez, and S. Ceniceros. 1960b. Selva de sombrerete (*Terminalia amazonica* (J. F. Gmell.) Exell y tres de sus fases sucesionales. *In* Resumenes de los trabajos presentados. I Congreso Mexicano de Botanica, Mexico.

Odum, E. P. 1969. The strategy of ecosystem development. Science 164: 262-270.

Pendleton, R. L. 1941. Some results of termite activity in Thailand soils. Thai Sci. Bull. 3: 24-53.

Popenoe, H. 1957. The influence of the shifting cultivation cycle on soil properties in Central America. Proc. 9th. Pacific Sci. Congr. 9: 72-77.

Richards, P. W. 1952. The tropical rain forest. Cambridge University Press, Cambridge, England. 450 p.

Rico, M. 1972. Estudio de la sucesión secundária en la estación de biología tropical "Los Tuxtlas". Tesis professional. Facultad de Ciencias, Univ. Nac. Autonoma de Mexico, Mexico, D. F. 28 p.

Rico, M., A. Gómez-Pompa, and J. A. Toledo. 1972. Uso de computadoras en investigacion de la sucesion secondaria en al Estacion de Biologia Tropical de la UNAM, p. 93-96. *In* El uso de computadoras en la flora de Veracruz. Inst. Biol., UNAM, Mexico, D. F.

Ross, R. 1954. Ecological studies on the rain forest of southern Nigeria. III. Secondary succession in the Shasha forest reserve. J. Ecol. 42: 259-282.

Sarukhán, J. 1964. Estudio sucesional de una area talada en Tuxtepec Oaxaca. Publ. Esp. Inst. Nac. Invest. Forest. Mexico. 3: 107-172.

Sarukhán, J., and A. Gomez-Pompa. 1963. Uso de tarjetas perforadas en la dinámica de la sucesión secundária, p. 18. *In* Resumenes de los trabajos. II Congreso Mexicano de Botanica San Luis Potosi, Mexico.

Schnell, R. 1971. Introducion a la phytogeogeographie de pays tropicaux, p. 552 to 559. *In* Vegetation et sols dans le monde tropical. Vol. 2.

Smith, A. C. 1973. Angiosperm evolution and the relationship of the floras of Africa and America, p. 49-61. *In* B. J. Meggers, E. S. Ayensu and W. D. Duckworth (eds.) Tropical forest ecosystems in Africa and South America: A comparative review. Smithsonian Institute, Washington, D. C.

Snedaker, S. C. 1970. Ecological studies on tropical moist forest succession in eastern lowland Guatemala. Unpublished Ph.D. Thesis, University of Florida, Gainesville. 131 p.

Snedaker, S. C., and J. F. Gamble. 1969. Compositional analysis of selected second-growth species from lowland Guatemala and Panama. BioScience 19: 536-538.

Sousa S., M. 1964. Estudio de la vegetación secundaria en la región de Tuxtepec Oaxaca. Publ. Esp. Inst. Nac. Invest. Forest México. 3: 91-105.
Stark, N. 1971. Nutrient cycling. I. Nutrient distribution in some Amazonian soils. Trop. Ecol. 12: 24-50.
Symington, C. F. 1933. The study of secondary growth on rain forest sites in Malaya. Malayan Forest. 2: 107-117.
Tergas, L. E. 1965. Correlation of nutrient availability in soil and uptake by native vegetation in the humid tropics. Unpublished M. S. Thesis, University of Florida, Gainesville. 63 p.
The Institute of Ecology. 1972. Man in the living environment. University of Wisconsin Press, Madison. 288 p.
UNESCO. 1960. Symposium on the impact of man on humid tropics vegetation. Goroka, UNESCO Publ. 402 p.
Vasconcelos, J. 1971. As regiões naturais do nordeste. Conselho de Desenvol. do Pernambuco, Recife, Pernambuco, Brasil. 422 p.
Vázquez-Yanes, C. 1973. Relación entre luz y germinación de semillas en algunas especies de la vegetación secundaria de una selva humeda, unpublished manuscript.
Vázquez-Yanes, C., and A. Gómez-Pompa. 1971. Estudios fisioecológicos de la germinación de semillas en algunas especies de la vegetación secundaria de una zona cálido húmeda de Mexico, p. 79. *In* IV Simposio Latinoamericano de Fisiologia Vegetal. Universidad Agraria La Molina. Lima, Perú.
Veloso, H. 1972. Vegetation of the Tucuri quadrangale. Quadrangale SA-22-ZC Projecto Radam, Belém, Pará, Brasil. Mimeo Report.
Vieira, L. S., W. H. dos Santos, I. C. Falesi, and J. P. Oliveira. 1967. Levantamento de reconhecimentos dos solos da região bragantina, Estado do Para. Pesquisa Agropecuaria, Brasil. 2: 1-63.
Wyatt-Smith, J. 1949. Regrowth in cleared areas. Malayan Forest. 12: 83-86.
Zinke, P. J. 1971. Preliminary legends for Radam. Projecto Radam Urca, Rio de Janeiro, Brasil. 35 p.
Zinke, P. J., S. Sabhasri, and P. Kundstadter. 1970. Soil fertility aspects of the lua forest fallow system of shifting cultivation, unpublished manuscript.

Section 5

INTERACTION OF MAN AND TROPICAL ENVIRONMENTS

Charles Bennett, Team Leader

Gerardo Budowski
Howard Daugherty
Larry Harris
John Milton
Hugh Popenoe
Nigel Smith
Victor Urrutia

Contributor: Enrique Beltrán

I. Introduction

Awareness that *Homo Sapiens* has had a significant impact on the ecosystems of the American tropics since his arrival on the continent several millenia ago is not new. Several studies have clearly documented that man has exerted major ecologic influences on neotropical environments dating back to the pre-Columbian period (e.g., Bennett, 1968; Sternberg, 1968; Denevan, 1970; Johannessen, 1963; Daugherty, 1974). Until recently, however, such awareness was confined to a small fraction of the scientific/scholarly community. Less than 15 years ago it was still appropriate to call for an appreciation of the ecological role that man has played in the American tropics (Sauer, 1958). It is a measure of recent progress that such a plea made to the ecological fraternity today would be largely unneeded.

However, this awareness has not been sufficient to generate a research effort that would provide critical data on what many feel to be among the most pressing ecological problems in the tropics. In the American tropics, these problems can be grouped into three broad categories: (a) an unprecedented increase in human numbers within a framework of limited natural resources, (b) a concomitant and unprecedented demand for food and other resources within the tropical region, and (c) a concomitant deterioration of the human environment to the extent that many feel that contemporary environmental problems are more critical in the developing countries of the tropics than in the more industrialized nations. These problems in the American tropics occur within an ecologic framework that is as yet inadequately understood. At least two major areas of prevailing ignorance are (a) the ecologic significance of a long history of environmental disturbance, vis-a-vis subsistence agriculture, commercial agriculture, livestock ranching, forest utilization, subsistence hunting, and more recently by hydrologic projects, transportation development, and tourism, and (b) the best strategies for management of the biotic and physical resources in tropical America. Such strategies involve resource utilization commensurate with the satisfaction of accelerating human needs and demands on the one hand and long-run ecosystem viability on the other.

The contemporary ecologic impact of man on neotropical environments entails the rapidly increasing uses of energy, materials, and space, which result in increases in productivity, expansion of communication and transportation networks, and increases in physical structures. These in turn allow an increase in human population and the level of living. Unfortunately, we also have numerous examples where mismanagement of resources has resulted in a decline in productivity, human carrying capacity, and standard of living. It is clear that man's impact on the land has definite thresholds, and that it is not clear how these thresholds are different in tropical regions compared to temperate and arctic regions. We also are not certain how local and regional environmental areas vary in their response to human impact within the tropics. This entire subject merits intense research effort since it is so closely tied to human well-being.

Human utilization of the land can be grouped into the following categories: agriculture, grazing, forestry, industry, urban development, transportation, and tourism. Each of these activities uses the environment in a special way, requiring certain inputs and resulting in specific outputs. Thus, the problem of analysis of the impact of man not only concerns the maintenance of a given system, but also the inputs and outputs of the system that link it to all other man-dominated and natural ecosystems. As populations are linked with other populations into communities and communities are coupled to other communities in the biosphere, similarly all human influences are coupled to other systems. For this reason, national boundaries may be economically and ecologically meaningless, as the activities of one country influence those of another through trade and the natural flows of wind and water. These regional linkages are discussed in Section 6. Here our attention is focused on the immediate impacts of man within the local rural and natural systems. On some, the major problems are identified; this chapter is not meant to be exhaustive. Rather, we have focused on the issues that especially merit the attention of ecologists concerned with the interactions of man and tropical environments. We have not considered the important impacts of industrialization and urbanization, both of which deserve special detailed study.

II. Recommendations for Research

1. Given the long history of human disturbance of the ecosystems of the American tropics and the increasing degree to which these same environments are being stressed by man, a most pressing research need is development of land-use planning and environmental-management methodologies that are not only ecologically sound but economically and socially viable as well. The innovation of such methodologies depends upon the collection of adequate data on and evaluation of the main environmental controls and determinants of agriculture, forestry, and other major human activities. Vitally important is the recognition that the existing data base for environmental parameters such as microclimatology, pedology, hydrology, biota, geomorphology, and geologic structure is insufficient for the development of ecologically sound land-management practices that are applicable to the American tropics.

2. Existing macroclimatic data are largely irrelevant for ecological, agricultural, and forestry problems. Urgently needed is a bioclimatic data base upon which to build land-use and environmental-management strategies. These data are fundamental to both basic and applied studies. The deficiency of bioclimatic data could be alleviated by (a) the extension and standardization of observing networks, instrumentation, and reporting services to enable a more precise description and interpretation of regional climatic phenomena and to provide a more solid climatological framework for agricultural feasibility studies, (b) the development of bioclimatological methodology and operational systems of observation applicable to the tropical ecosystems of Latin American and the innovation of bioclimatic models that are relevant to ecology, agronomy, planning, and biogeography, and (c) the initiation of microclimatic and energy moisture balance research as part of an overall program of ecosystem analysis in the neotropics in order to resolve fundamental problems in ecology, biogeography, and atmospheric energy and moisture transfers. Such research would alleviate the basic problem of data efficiency from the earth/atmosphere interface—the vertical zone most important for ecology, agronomy, and biogeography.

Equally important is the collection of basic data on soil, water, biota and other important environmental parameters.

3. As the world food problem caused by rapid population growth becomes increasingly severe, more attention is being focused on the agricultural potential of the tropics. Recent successes with the development of new varieties of many basic grains have created a great optimism, but it is quite apparent that any major improvements in agriculture will require a large sustained research effort to continue the development of new varieties as pests and diseases evolve overcoming the resistance of established varieties.

The environmental controls of agricultural production in the tropics should be studied utilizing the basic information available in plant and animal physiology and ecology. Especially significant research areas are the following: (a) microclimatic effects on organisms and processes, (b) manipulation of plant nutrients, (c) effects of different land-use systems on soil-fertility structure, (d) the genesis of soil structure in different tropical soils, (e) the role of mycorrhiza in tropical nutrient cycles, (f) crop combinations that simulate the natural multistrata, species-rich communities found in the tropics, (g) the application of systems analysis to traditional agricultural systems, multiple cropping systems, and integrated agricultural systems, and (h) methodologies to combat agriculturally induced soil erosion.

4. Careful attention must also be given to a number of research possibilities that would provide information basic to an understanding of shifting cultivation and other types of peasant agriculture. Research priorities include (a) study and development of modifications of shifting cultivation to increase productivity, e.g., by fallow improvement through planted successions, the use of legumes as cover crops, multiple cropping, and multistory cropping, (b) the effects of shifting cultivation on soil structure and nutrient levels, (c) factors controlling the length of fallow necessary to avoid ecological deterioration of the land, (d) yields under shifting cultivation with various combinations of inputs, (e) effects of fire and deforestation on soil micro- and macrofauna, (f) temperature profiles during burning and the variables influencing fire behavior (including fuel size, type, quantity, and spatial arrangement, fuel moisture, topography, and wind velocity and other atmospheric conditions), and (g) influence of extensive fires and deforestation on rural air pollution and regional climates.

5. Traditional agricultural systems, multiple cropping systems, integrated agricultural operations using plants and animals should be investigated utilizing modern techniques of systems analysis when costs and benefits from both environmental, economic, and social points of view are considered.

6. The forests of the American tropics have experienced a long and unfortunate history of clearance, disturbance, and utilization by man with poorly understood ecologic and economic consequences. A variety of tropical forestry-utilization practices has been proposed, but they have yet to be evaluated objectively, taking into account the economic, ecologic, and social costs. In particular the following forestry practices appear worth investigating: (a) improving harvesting and management schemes for various types of secondary forests, which are generally more responsive to silvicultural treatments resulting in sustained yield productivity, including selective cuttings and schemes such as shelterwood, enrichment, and plantings, to improve the quality and regeneration of the stand, (b) establishing man-made plantations in mosaics of area, space , or age groups, to minimize

the danger of insect and disease epidemics and destructive influences that may be associated with large tracts of single species, (c) intensifying or concentrating sustained yield management practices in those natural forests that contain a large proportion of one or a few harvestable species that lend themselves to successful regeneration schemes (e.g., *Virola, Carapa, Prioria, Dialanthera, Mora,* mangroves, and tropical pines), (d) managing tropical mixed forests by using refining techniques, enrichment schemes, and other practices that are feasible under the prevailing economic conditions, and (e) replacing subsistance agriculture on forest land with a permanent forest industry.

In addition, other major research topics *vis-a-vis* tropical forestry are (a) the location, classes, and condition of natural forests, (b) alternative sources of fuel (in view of the fact that firewood and charcoal constitute major sources of household and industrial energy supplies throughout the American tropics, (c) the advantages and disadvantages of large companies with large capital investment in comparison with small enterprises with limited capital regarding such factors as sustained productivity, influence on nearby natural forests reserved for noncommercial timber production, and quality of life of those dependent upon the industry, (d) the conversion of the high productivity of leaves and wood into edible products (for man or domesticated animals) through physical-chemical processes, (e) forest-management strategies for recreation, scientific purposes, the preservation of valuable species and gene pools, and protective purposes (for water, soil, wildlife, etc.) as well as management for timber production, and (f) conservation attitudes toward and public awareness of natural forests and their potential in promoting development.

7. Similarly, the faunal resources of tropical America have been seriously depleted with little understanding of the long-range economic, ecologic, and esthetic implications. Major research problems are (a) the ecologic roles that individual taxa play in ecosystem structure and function, (b) an evaluation of the resource value of tropical animals as sources of food, recreation, and gene pools of potential human use, (c) identification of taxa that could be ecologically managed on a sustained-yield basis, and (d) investigation into the possibilities of domestication or semidomestication of tropical animals for human benefit.

The scientific types of data needed to resolve the above research questions are (a) the present status of individual animal taxa, particularly those of potential use to man as a food resource, (b) the maximum sustainable cropping rate of individual species, (c) accurate life-history data on exploited taxa, particularly concerning reproductive behavior, (d) identification of animal taxa that could be effectively managed ecologically, (e) the nutritive value of different animals, (f) the impact of pollution on the quantity and quality of freshwater and estuarine fishes, (g) the impact of certain fishing technologies (such as the use of dynamite and plant poisons) on productivity, (h) cultural preferences and taboos regarding the use of wild animals, (i) the potential of wildlife management to provide high-quality protein for human needs, and (j) census data on commercially exploited taxa.

8. The protection of the faunal resources of the American tropics is based on several unanswered research problems, including (a) documentation of which native animal taxa are threatened, endangered, or exterminated, (b) documentation of which habitats and environments are threatened or endangered, (c) identification of the principal causal factors (past and present) accounting for the depletion of specific animal taxa, (d) identification of spatial and other habitat requirements of endangered species including a precise determination of the minimal area required for the maintenance of an ecologically viable wildlife refuge, and (e) the development of research methodology to demonstrate more concretely the economic, ecologic, cultural, and scientific values of endangered species and environments.

9. Wildlife management for tourism requires information in the following areas: (a) an evaluation of the potential of the American tropics to support a wildlife-based tourist industry, (b) an identification of the historical, contemporary, and expected values of wildlife manifest in the social mores and institutional structures of the societies in question, (c) the role of wildlife in urban, regional, and national planning based on utilitarian concepts as well as ecologic and esthetic values, (d) the degree and extent of natural wildlife communities necessary for the sustained integrity of viable human populations, and (e) the adequacy of legislative statutes and the degree to which they are enforced.

10. There are numerous examples documenting the detrimental ecologic, epidemiologic and sociologic impacts of hydrologic projects in the Asian and African tropics. The increasing degree to which river basins in the American tropics are being regulated and modified demands research into the impacts of dams and reservoirs on river-basin ecology, aquatic flora and fauna, human resettlement, the spread of disease organisms, accelerated dam siltation, accelerated erosion downstream, and accelerated coastal sedimentation. Until thorough scientific investigations of the environmental and social costs and benefits in river-basin development are included as an integral part of development planning, unforeseen and costly perturbations are likely to occur.

Specifically, the major research questions to be addressed are (a) the impact of hydrologic projects on freshwater fish ecology, particularly on migration, feeding, and spawning habits, (b) the dynamics of nutrient supply, temperature, and other factors influencing reservoir productivity, (c) the role of riverine fish populations in the diet of river basin cultures, (d) the potential value of aquaculture in newly created reservoirs, (e) the potential value of mixed irrigated agriculture and fish systems, (f) the ecologic impact of introductions of exotics, (g) the ecology of schistosomiasis and other water-borne diseases, (h) biological and mechanical control of aquatic weeds, (i) the potential utilization of aquatic weeds as resources, e.g., as fodder and fertilizer, (j) the interrelationships between rainfall, soils, vegetation, and alternative land-use patterns in river basins undergoing development or determine optimal land-use methods and management strategies, (k) the impact of reservoir sedimentation on loss of storage capacity, reservoir life, lake recreation, fishing, and other lake resources, (l) the downstream impacts of reduced silt loads and nutrient flows, (m) a survey and inventory of the value of natural areas, scenic sites, and the biotic, scientific, and historical resources likely to be inundated, (n) identification of areas prone to salinization, and (o) the effects of salt contamination of downstream freshwater on agricultural production and aquatic biota.

11. Research is needed on the impacts of mining, transportation development and tourism in the tropical environments. The following research questions are identified: (a) the ecologic impact of historical and contemporary mining practices on the landscape, particularly in terms of inducing permanent alterations of the vegetation cover, (b) the ecologic impact of oil spillage on tropical terrestrial as well as marine ecosystems, (c) the ecologic, epidemiologic, and sociologic impacts of transportation development, particularly in terms of highway construction through mature rainforest areas, (d) the degree of ecologic deterioration of existing parks and reserves through misuse and mismanagement, (e) the location of new parks and reserves to disperse the tourist impact, (f) an ecologic determination of human carrying capacities for tropical park environments, and (g) the innovation of planning and management techniques to prevent misuse and overuse of parks and reserves.

III. Evaluation of Ecological Determinants

Much of the economic development throughout the American tropics has been achieved only at a significant environmental cost. If there is any lesson to be learned from conservation ecology, it is that economic development and progress can only be sustained within a framework of regional environmental diversity and stability, i.e., short-term economic gain at the cost of regional environmental stability is self-defeating in the long-run. At present there is a clear lack of an adequate conceptual base upon which to base sound, sustained development plans in the American tropics that would optimize economic benefits and environmental stability.

A critical need in the development of sound environmental-management practices is an inventory of the main environmental determinates of agriculture, forestry, industry, and other activities. Such an inventory includes a description of bioclimate, biota, hydrology, topography, and soils. Several of these features have been discussed in earlier sections since they influence nonhuman population and communities as well. Here we will focus on bioclimatology.

Fields of climatology relevant to a consideration of the interrelationships between man, climate, and ecology are (a) macroclimatology, which includes descriptive, dynamic, and energy balance climatology, (b) bioclimatology or microclimatology, and (c) climatic change and weather modification. Although this discussion focuses upon microclimatology, a few comments on the research status of macroclimatology are in order lest the misconception arise that there are no remaining research problems in this larger-scale field of investigation (also see Section 6).

The lack of extensive, reliable, and comparative data seriously limits research in macroclimatology in Latin America. The distribution of meteorologic observing sites throughout most of Latin America is closely related to the distribution of human population. Thus, extensive land areas, particularly the tropical lowlands, are largely unobserved meteorologically. These ground stations are especially needed to establish "ground truth" for satellite climate observations.

Meteorologic observation of adjacent marine areas which have a pronounced influence on terrestrial climates is largely restricted to discontinuous observations aboard ships which transverse only a small portion of the oceanic sectors influencing climate in the Latin American tropics. Such observations are not monitoring conditions at or near the air/sea interface, but are measuring on-deck conditions highly modified by the vessel itself. Here again accurate "truth" measures of sea conditions are also required for total-climate studies.

Other serious limitations imposed on research in tropical macroclimatology include the limited number of observed parameters (mostly confined to precipitation and temperature), instrument error, methodological error, and the lack of standardization in observing station construction, instrumentation, and exposure. Data problems *vis-a-vis* climatology in Latin America have been reviewed by Daugherty (1971).

Standard weather data have more serious limitations for research in bioclimatology, the field of study that treats those microclimatic phenomena that are directly interrelated with biological and ecological phenomena. Standard observation stations are not strategically located with reference to the natural structure of ecosystems; rather the observations are generally made above an artificial grass surface in open and disturbed sites. The data collected from such sites are rarely applicable to biogeographical and ecological problems such as the distribution of plants and animals, the nature of animal habitats, and ecological energetics nor are they applicable to analogous human activities. Air temperature measured at the standard observational level of 1.5 m, for example, is not a particularly important biological datum. Numerous measurements have documented the large lapse rates that occur within the first 2m of the ground, particularly within a few millimeters of the surface—the interface to which most biological life is exposed. Observations of the true heat stress on tropical organisms are obviously lacking.

In sum, standard weather observations are largely irrelevant for biological and ecological problems and provide little insight into the energy conditions at the earth/atmosphere interface—the key element of bioclimatology. Thus research in bioclimatology and the related fields of ecology and biogeography must rely on a distinct data base. Such a data base, however, is extremely limited throughout the American tropics. Attempts to measure the vertical structure of standard climatic parameters have been short term, haphazard, and scattered. The systematic observation of energy- and moisture-balance parameters such as solar and thermal radiation, albedo, evaporation, and the fluxes of latent heat and soil heat has only recently begun at a few localities in spite of their importance in ecology, agriculture, and meteorology. Observations of important agronomic parameters such as evapotranspiration, soil temperature, net photosynthesis, and carbon dioxide exchange are almost nonexistent.

In spite of data limitations, there have been attempts to construct "bioclimatic systems" utilizing standard weather data, many of which have been used for planning and development purposes. Nevertheless, each of these systems has been shown to have serious deficiencies. Clearly there is a critical need for the innovation of bioclimatic models that have more relevance for biological distributions, crop plant tolerances, and agricultural development in general. However, such innovation must await the initiation of the collection of data that are ecologically relevant.

The knowledge of the vertical structure of climatic parameters within different ecosystems in Latin America is limited. Richards (1952) has summarized much of the knowledge of tropical forest microclimate. An additional important source of forest microclimatic data is Schulz's work (1960) on rain-forest ecology in Surinam, which included the measurement and analysis of the vertical structure of light, atmospheric humidity, evaporation, air temperature, and soil temperature.

There is a critical need for the measurement of energy and moisture balances of the various ecosystems of Latin America. Although certain energy and moisture parameters are measured at a few stations, there has been no systems approach to the surface observation of energy balance, with the exception of a recently initiated program in

Barbados, which has limited relevance to tropical mainland areas. Indeed, no standard systems methodology has yet been devised for the measurement and analysis of energy and moisture balance parameters. This lack of energy/moisture systems data from Latin America precludes a precise understanding of many ecological problems such as energetics and biogeographical distributions, adaptations, and tolerances. The lack of surface data also hinders an understanding of the importance of the lower latitudes in larger scale atmospheric transfer of heat and moisture.

More extensive and precise observation of solar radiation—the key factor governing photosynthesis—is essential for estimating potential agricultural productivity in the tropics. Chang (1970) has estimated that potential photosynthesis during a 4 month (high sun) growing season in the humid tropics is approximately two-thirds that of the middle latitudes. However, mean annual photosynthesis is higher in the tropics than in the middle latitudes and is higher in the highland tropics than in the lowland tropics. In addition Borden (1941) has shown that fertilization is most effective under conditions of high solar radiation and that the use of fertilizers is often uneconomical in areas of high cloudiness (such as the humid tropics).

Thus, it is becoming increasingly apparent that agricultural production in the tropics might be increased through the intensification of agriculture in the uplands (by double cropping, irrigation, and fertilization) rather than by expansion into the lowlands. It may be that the ecological functions of the lowland tropical rain forests most beneficial to man are not agricultural, but rather the provision of animal habitats and the stabilization of watersheds and regional hydrologic cycles. The determination of optimal ecological, economic, and social uses is a pressing research task of the ecologist and would aid the proper management and utilization of the tropical ecosystems in Latin America that are under increasing pressure by man.

The systems approach offers the most fruitful and productive research technique. The role of the bioclimatologist in ecosystems research is clear: the observation and analysis of microclimatic parameters that are ecologically and biologically relevant. Ideally, systems research would be conducted in at least one permanent study site in each major ecosystem in the tropics. Such sites should be equipped for the continual measurement of (a) the vertical gradients of air temperature at several levels, soil temperature, moisture, and wind, (b) the energy balance (solar and thermal radiation, albedo, and the fluxes of latent heat, sensible heat, and soil heat), and (c) the moisture balance (precipitation, evaporation, absorption, and runoff). Study sites should also have the instrument capability for shorter term observation of the component biotopes (or microenvironments) of the ecosystems in question. The minimal instrumentation required for such observational sites is (a) two net-radiation recording systems,

(b) two pyranometers, (c) one infrared radiometer, (d) one six-thermistor temperature recording system, (e) three microhygrographs. The approximate cost of the above instruments is US $7,500 in addition to the construction and equipment for a first-order standard weather station of approximately US $1,000.

The data thus collected would be applicable to a host of research problems in (a) ecology—such as energetics and productivity, (b) agronomy—crop tolerances, ranges, and productivity, (c) energy systems —local, regional, global, and (d) biogeography—plant and animal distributions and the nature of habitats. Such data would contribute greatly to an understanding of the structure and function of ecosystems —an understanding of which is vital for the formulation of alternative strategies for ecosystem utilization and environmental management —basic problems in applied human ecology. Specific examples of the applicability of bioclimatic data to agricultural development are given in Part IV.C. of this section. Many of the research techniques and tools of bioclimatology and ecology have been field tested. Critically lacking are operational systems for observation and analysis.

The most feasible and potentially productive avenues for increasing the output of research are through the strengthening of interdisciplinary graduate training in climatology and ecology (which is woefully lacking in American universities) and the establishment of permanent research stations in tropical America for ecosystem analysis.

In addition to bioclimatic data, information on other basic environmental components, e.g., soils, hydrology, biota, geomorphology, and geologic structure, must be incorporated into the data base for environmental-management strategies.

IV. Agriculture and Livestock

The tropics contain many characteristics that can contribute to highly productive agricultural systems, and indeed most of the record yields in terms of calories per hectare per year occur in tropical climates. On the other hand, the intensive development of tropical agricultural systems over vast areas has lagged considerably and some limiting factors appear to be quite formidable. Sir Dudley Stamp pessimistically observed: "the tropics are a region of lasting difficulty." Tropical landscapes in many areas are going through very rapid periods of change, and an analysis of some of the environmental factors contributing to the relative success or failure of new developments is essential. Furthermore, the use of ecological principles and the development of environmentally- oriented research should increase the prospects of viable agricultural methods which should be productive over long periods of time.

Some obvious advantages for plant growth in the humid tropics of America are a high annual amount of solar energy, adequate supplies of moisture, much available space for expansion, a growing season of 12 months and well-structured soils. Although these factors contribute to much of the high biological productivity or biomass in many tropical regions, they may impose serious obstacles to the development of rational agricultural systems.

The lack of large seasonal variations in the humid tropics seriously complicate the problems of pest and weed control as mentioned earlier. Means must be found to break life cycles or minimize competing populations. The diversity of the tropical biota further complicates the pest problem. The high-moisture regime coupled with the geologic stability of vast regions has produced a soil low in nutrients and nutrient retention. The total amount of precipitation in the tropics is generally much higher than in temperate regions. Certainly related to the development of tropical agriculture is the effect of climate on such factors as human health and comfort.

A. Research on Agricultural Systems

The following discussion of agricultural systems is not intended to be comprehensive. Instead, emphasis is placed on those systems that are strongly affected by environmental fluctuations or where significant improvements could be made by incorporanting new modifications to increase productivity and minimize the need for industrial inputs. Such agricultural types as plantation systems, rice agriculture, row crops, or other intensive systems of monoculture are excluded. Obviously, much of the potential research described in the preceding section would make significant contributions to the improvement of many different systems.

"Slash and burn" or shifting cultivation is an agricultural system in which tracts of forest are cleared by cutting or girdling and firing, cropped for short periods of time, and abandoned. It represents the simplest form of a land-use system and, as practiced by people in sparsely populated areas of the tropics, causes relatively few long-range ecological disturbances provided the cropping/fallow ratio is maintained at an appropriate level and is practiced on gentle slopes. The destruction of forest and use of fire are possibly the two main characteristics of this system most likely to induce ecological disturbances. Several factors can control the extent of these ecological disturbances, e.g., length of cropping in relation to the fallow, topography, intensity and frequency of fires, vegetation type, and climatic conditions.

When a forest is burned, nutrient cations and phosphorus from the vegetation are accumulated in a thin layer of ash on the surface soil, resulting in a temporary increase in the fertility of the soil.

This effect is generally accompanied by a rise in pH. Data from Africa (Nye and Greenland, 1960), for example, show pH increases ranging from 0.7 to 2.7 following burning of 10 to 40-year-old secondary forest. Changes in soil reaction and rate of return to the original pH following burning will depend on many factors, such as type and amount of ash, soil characteristics (buffering capacity, original pH, permeability to water, texture, etc.), intensity and distribution of rainfall, and land management during the cropping period. This rise in pH can constitute an important ecological factor, particularly in acidic tropical soils with poor buffering capacity. An increase in pH from 4.5 to 5.5, for example, could completely neutralize the toxic effects of aluminum in the soil, markedly accelerate the rate of decomposition of organic matter, and significantly alter other microbiological processes in the soil such as nitrogen fixation, nitrification, denitrification, and mycorrhizae development. With the exception of nitrogen and sulphur, burning apparently does not result in direct losses of nutrients from the ecosystem. However, most nutrients are immediately made available for plants at the soil surface, but can be lost by leaching or erosion.

The population and composition of soil microflora change after forest clearing and burning due to the decrease in soil acidity, increased supply of nutrients, changes in the amounts and composition of soil organic matter, and removal of the vegetation cover, which results in increased surface temperatures, accelerated surface runoff, and evaporation, and decreased moisture in the surface soil. However, reports of observed changes in microbial populations following burning have generally shown an increase in the nitrifier organisms. This effect could be attributed primarily to the increased pH and increased supply of nutrients in the soil. Changes in soil microbial populations due to burning are generally followed by a gradual return to the original level.

The forest floor litter contains a diverse mixture of microfauna, viz., earthworms, ants, termites, centipedes, millipedes, molluscs, and other invertebrate organisms. On clearing and burning, forest litter is destroyed and the soil is left and exposed to the sun until the subsequent crop forms an effective cover. These changes in the soil microclimate have a marked effect on the soil microfauna. The temperature of the surface soil may rise above the temperature tolerated by most soil organisms and their activity ceases (Ewel, 1971). Studies on this subject, however, are scarce under tropical conditions.

The surface soil under forest fallows generally exhibits a crumb structure and a high degree of porosity due to its high organic matter content and the activity of the soil microfauna. Burning destroys the layer of nonhumified organic matter and the soil is left bare and unprotected to the mechanical action of rains. The extent of structural deterioration of this soil layer depends on soil type, prevalent microclimatic conditions, topography, and types of cropping and soil-management systems. Many tropical soils, e.g., Oxisols, have a strong-structure stability due to the effect of iron and aluminum sesquioxides, and therefore these soils can generally conserve their physical conditions under frequent cropping. However, much more needs to be learned about changes in soil structure under conditions shifting cultivation.

Many tropical soils are sufficiently stable to resist erosion for a few weeks until the crop cover develops. Plots are cleared in patches that are surrounded by forest fallow, reducing the risk of runoff losses. Another factor that helps to reduce erosion is the common practice of the shifting cultivator not to disturb the soil by cultivation. During the first year after clearing a mature forest, erosion losses are generally insignificant. However, burning and cropping during subsequent years increases the risk of erosion because hoeing and cutting is required to control weeds and because crops take longer to form a protective cover over the soil. The use of the machete to rid the land of weeds, as widely practiced in tropical America, constitutes a method of reducing soil loss in these cases.

In soils with ground water close to the surface, forest destruction may cause a rise in the water table, resulting in the production of swamp conditions. Although not common in soils with plinthite at or near the surface, the removal of forest can result in the irreversible hardening of the soil. In both cases these of, the return of these lands to forest or agricultural productivity would be difficult.

Shifting cultivation has stood the test of time under conditions of low population densities. This agricultural system is practiced by people on all three continents with tropical areas. It now appears that it is the best system that could have been devised for subsistence farming in low-populated areas. The ecological impact of this system is minimized by the use of short cropping periods and sufficiently long fallows. However, as human population increases and land becomes scarcer, the peasant will have to clear land fallowed for shorter periods and increase the frequency of fires, which will undoubtedly result in deterioration of the environment.

The tropical regions permit year-round plant growth if water conservation is practiced. The unused land available for agriculture is becoming scarcer and much of this land lacks appropriate infrastructure, is of poor fertility, and does not have irrigation or drainage facilities. Therefore, a major potential lies in the more intensive use of the land now under cultivation.

Effective multiple cropping minimizes the number of days the available agricultural land is unused. Research on multiple cropping agricultural systems has been conducted mostly at the International Rice Research Institute (IRRI). The IRRI research has concentrated on rice, soybeans, sweet corn, grain sorghum, and sweet potatoes. These crops are planted in rotation, overlapping their growth cycles to some extent. Rice can be planted as the first crop in early June at the beginning of the rainy season. Sweet potatoes can be planted as a second crop 10 to 20 days before harvesting the rice, when any surplus of water remaining on the soil surface has been drained off. This phase of the rotation thus saves 10 to 20 days. A similar approach is used to plant soybeans, sweet corn, soybeans for vegetables, and rice which constitutes the end of the annual cycle. In the Philippines, good crops of all these plants have been grown in one calendar year on the same land.

Multiple cropping is flexible and modifications can be made varying crop sequences according to price or demand. This system should be tested in the American tropics with different rotational schemes and management systems based upon the economic, ecologic, and social conditions of the region. According to Bradfield (1969), as much as 22,000 kg ha^{-1} of food grains have been produced at the IRRI in 12 months, a feat that cannot be duplicated in the temperate parts of the world. This system not only makes it possible to grow much more food, but it can also provide the tropical populations with a balanced diet. Multiple cropping also reduces fertilizer leaching, providing a more complete cover on the soil, resulting in less erosion. It could also reduce insect and pest attacks by interrupting competition of life cycles and thus would require less use of chemical pesticides and herbicides; and it would maximize the utilization and conservation of rainfall or irrigation and sun energy.

Since the Spanish Conquest, cattle raising has been a widespread agricultural use of land in the American tropics. The Spaniards brought a tradition of livestock raising, and today, throughout much of Latin America, ranching is a way of life even though economically it may not be highly remunerative in many regions. However, the use of livestock will continue to be important on marginal soils that are neither suitable for intensive agriculture or forestry. Indeed, livestock production will probably grow steadily in the future as world markets expand and management practices improve. Livestock production in Latin American now receives more assistance from international loans than any other agricultural enterprise. Several basic management tools deserve the attention of the ecologically-oriented investigator.

The use of fire for pasture management is widespread and is an important management tool for the elimination of unpalatable growth, for control of regeneration of woody growth, for the destruction of ticks and snakes, and reduction of the hazards of spontaneous fires. The effects of fire can be greatly moderated by the time of burning during the day, amount and direction of wind, dryness of vegetation, and topography. Manipulation of these factors can produce a cool or hot fire and some experienced practitioners can tailor-make the vegetation by judicious burning. One research priority should be to relate type and timing of fire to regeneration and vegetational succession. For example, one could time a fire to eliminate some undesirable plants before seeding and encourage others.

One obvious improvement for immediately and cheaply improving extensive livestock systems is the use of mineral supplements, which are not widely used in tropical America. A correlation of mineral deficiencies with geological and soil factors would assist in the formulation of mineral supplements that are specific for different localities.

The use of legumes in tropical pastures will be discussed in the section on biological nitrogen fixation. At present, no really good legume/grass mixtures exist for the humid tropics. The legumes either cannot compete with the grass or eliminate it by shading. Many other useful legumes can be quickly eliminated by overgrazing. Research on a desirable species- mix and the necessary management practices including burning, grazing, interval, mowing, and the use of nutrient supplements with fertilizers is greatly needed.

Other obvious practices for improved livestock production would be the use of a managed breeding and calving season to coincide with dry periods, energy and urea supplementation during seasons of poor pasture growth, and the elimination of cows that calve less frequently. Detailed work on the ecology and suppression of certain livestock diseases would also be useful (see Section 2 for a discussion of livestock disease problems).

In lowland areas of the southeast Asian tropics where agricultural population densities have been high for centuries, many useful systems have developed that closely integrate plants and animals. These systems are usually much more efficient than simpler types of agriculture because maximum use is made of nutrients through recycling and because they are more diverse. Many of these systems might be adapted to tropical America, but cultural attitudes must be considered when implementing such changes.

An obvious representative type of this form of food production is the use of plant by-products to feed livestock and the use of the resulting manure to fertilize crops. A more complex system involves the use of plant wastes to feed livestock that are placed in pens over fish ponds. Waste from the harvested fish or water from the ponds would be used to fertilize crops. Although these practices if not carefully controlled could also promote the cycling of diseases or pests, they represent highly efficient uses of space and nutrients. Much attention is being paid to these types of practices in many industrialized countries where by-products of intensive agricultural systems have created serious problems of pollution or waste disposal.

Ecological research on mixed crop-animal systems should prove to be highly productive. Likewise research could be fruitful on alternative energy pathways in such systems. Inputs could also be introduced at several different points in the cycle to improve the productivity of desired foods. The use of different combinations of species may also increase efficiencies of food production. Further, research on physical environmental factors influencing the process are essential to develop these mixed systems.

B. Conservation of Soil Structure

Agricultural systems differ in numerous characteristics, which may improve, maintain, or deteriorate soil structure. These include the type of root system (biomass and distribution), ability to conserve soil organic matter, moisture requirements, tilling and mechanical cultivation requirements, extent to which the soil is unprotected from rain, and fertilizer requirements. On the other hand, the effects of these differences vary with the soil, particularly in relation to the nature of the mechanism by which soil aggregates are stabilized. It appears that the biotic activity in the soil is the major factor responsible for the production of soil aggregates and their stabilization. The effectiveness of earthworms and termites in reworking the soil is well known (e.g., Barley, 1961). Other components of the soil macrofauna such as ants, centipedes, springtails, and beetles, have a similar effect. Cementation of soil particles into stable aggregates in many soils is also due to organic matter (Bradfield and Miller, 1954). These polymers are strongly attracted by clays and act by adsorbing to the walls of soil pores in the two to 50-micron size range. This results in considerable strengthening and resistance to dispersion.

In addition to polysaccharides and other organic polymers, inorganic materials can also affect cementation of soil particles, especially in the tropics. Calcium carbonate, silica, iron, and aluminum hydrous oxides are in this category. In subhumid to arid regions, calcium carbonate and silica are the most common cementing agents. According to Greenland (1971), soils in South Australia containing calcium carbonate have been shown to have aggregates whose stability is little affected by removal of organic matter. In most tropical soils, e.g., Oxisols and Ultisols, the iron and aluminum hydrous oxides seem to play a more important role on aggregate stabilization than organic matter.

The deterioration of soil structure may occur through the effect of many agricultural practices, such as liming, tilling, fertilizer application, pesticide use, and soil compaction by cattle or farm machinery. It is of the utmost importance, therefore, to study the genesis of soil structure in different tropical soils and its conservation under different agricultural management systems.

C. Climatic Effects on Organisms and Processes

Solar radiation impinging on tropical rain forests has its maximum absorption in the top layer of leaves and branches. As the radiation filters through the strata of leaves the intensity decreases and spectral changes occur. Plants in lower strata may receive much less light but utilize it much more efficiently for photosynthesis than the leaves exposed to full sunlight. Vertical temperature and moisture gradients similarly occur within the strata of rain-forest vegetation. Several types of agriculture widely used in Southeast Asia take advantage of this vertical stratification of energy relations and microclimatic conditions. These systems have developed over a long period of time. In these systems, where annual crops are planted beneath perennial crops, one often finds papayas or bananas growing under coconut palms. Pineapples, eggplants, or other low growing crops may be planted in spots that are not excessively shaded. This stratified system has a twofold advantage: (a) a certain element of diversity and a maximum utilization of space and (b) the achievement of productive potential per plot of land. Some house gardens also are characterized by this type of structure.

Research on the microclimates under such crop systems and a determination of climatic requirements of different crop plants in response to variation in temperature, moisture, light, and soils is needed. These data would help in the optimization of agricultural production in different environments. We expect that land-use systems that simulate the natural vegetation should be well adapted to the tropics. For example, many regions of moderate to high elevation within the geographical tropics are characterized by cool nights and warm days throughout the year. Since the release of inorganic forms of nitrogen and sulfur for plant uptake from organic matter is a function of microbial activity, the availability of these nutrients is greatly diminished by low night temperatures. Thus, many areas of the high Andes and Central American highlands are quite deficient in nitrate and sulfate for crop use, even though soil reserves may be high. Most plants only take up these two elements in the most oxidized form. Research should be undertaken to ameliorate this problem of high altitude agriculture in the tropics.

As a first step, an ecological study to establish the correlation between temperatures and sulfate and nitrate release should be made. These data could then be used to map out ecological zones or to determine vegetation types where such problems of nutrient deficiency should be expected. Actual field trials then could determine or refine the predictive accuracy of these methods. Perhaps research with mycorrhizae at high altitudes might lead to management methods whereby plants nutrient uptake would occur directly from organic matter or organic fertilizers without the need of other forms of biological transformations.

Little applied knowledge is available on heat stress in animals and plants in the humid tropics. The high atmospheric humidity produces a much greater heat load on organisms. Leafy vegetables wilt in the middle of the day though soil moisture may be adequate. Livestock seek shade instead of grazing, and dairy production is lower. More information is needed on the physiological responses to this environmental stress. Applied research on management methods to minimize production losses would also be highly useful. One obviously promising avenue of investigation is the correlation of nutrition with heat tolerance. Additional work on the correlation of surface color and pigmentation with ambient temperature stress would be useful.

The amount of available soil moisture at seedling stage is an important agricultural problem in the wet-dry tropics. At high latitudes crops are usually planted in the spring in soils soaked by the thaw of winter snows. An abundance of moisture is initially present, temperatures are cool, and seedlings are not subjected to serious moisture stress. In contrast, the planting of crops in the wet-dry tropics usually takes place at a hot time of the year at the onset of the rainy season. Soils are thoroughly dry with no moisture reserves. During initial growth of the seedling, rainfall is quite sporadic and the young plants may go through several periods of moisture stress. Planting at a later date is usually inadvisable because weed and pest populations become established. Furthermore, nitrates are released very rapidly from organic matter at the beginning of the rainy season and reserves are seriously depleted later. If crops are not present to take advantage of this early nitrate release, much of the nitrogen will probably be lost by leaching. Research to alleviate the problem of early moisture stress should include basic physiological research on mechanisms of resistance, which could be incorporated into breeding programs. Other promising research should include studies of the depth of planting, mulches, cultivation, and the effects of certain nutrients on plant growth.

D. Plant Protection

Perhaps one of the most serious though often overlooked problems in humid tropical agriculture is the extreme difficulty of plant pest and weed control. Without cold winters or dry seasons pest populations increase rapidly. Species diversity in the tropics often makes it difficult to focus on a few target organisms. Forms of integrated controls will become more important in the future but persistent broad-spectrum pesticides will probably continue to be used for a long time. Many tropical peoples remember the millions of lives that have been saved by malaria control through DDT, or the vast areas of land that are now habitable for the same reason. However, one must be extremely apprehensive when we remember that many of these same countries, with little analytical capability for determining pesticide residues, will be confronted with very stringent restrictions on their export crops in certain foreign markets—a dilemma of major proportions.

Therefore, much of the research on pesticides and biological controls now underway in the temperate countries assumes even greater significance in the tropics. Research on biological controls and highly specific, quickly degradable insecticides for target populations is important, as well as increased knowledge of the mechanisms of insect and disease resistance to chemicals. A much more elaborate statement of some of the problems and implications of pest control is discussed in Section 2.

E. Biological Nitrogen Fixation

Nitrogen is generally the most deficient plant nutrient in tropical soils. The use of inorganic nitrogen fertilizers is not feasible in large areas of tropical regions due to their low efficiency and the high cost of forms now available. Furthermore, prolonged use of some inorganic forms could result in marked soils acidification, aluminum toxicity, accelerated leaching, losses of nutrient cations, and breakdown of soil structure.

The most important form of symbiotic nitrogen fixation is that resulting from association of legumes with *Rhizobium*. Tropical legumes have been reported to fix nitrogen in amounts ranging between 50 to 250 kg $ha^{-1}yr^{-1}$ (Henzell and Norris, 1962). Cultivation of legumes in the tropics, therefore, could constitute an effective method of adding nitrogen to the soil. Furthermore, this nitrogen is more effective than adding an equivalent amount of inorganic nitrogen fertilizer and it would cause less soil deterioration.

Legumes are generally considered to have higher calcium requirements than cereals or grasses. Tropical legumes, however, are naturally adapted to acid soils of low fertility and require less bases and phosphorus from the soils for effective nodulation. Differences in sensitivity to nutrient deficiencies between tropical and temperate legumes appear to be related to a greater ability of tropical legumes to extract their nutrients from the soil rather than to differences in their requirements.

Nitrogen fixation has also been reported or implicated in other plant groups: (a) symbiotic nitrogen fixation in root nodules by nonlegumes has been reported in at least nine families—four from tropical regions, viz., Zygophyllaceae, Podocapinae, Cycadaceae, and Casuarinaceae, (b) some genera of the families Dioscoreaceae, Rubiaceae, and Myrsinaceae have cavities within leaf tissues occupied by phyllosphere bacteria, which may fix nitrogen (Silver et al., 1963), (c) nonsymbiotic nitrogen fixation in the soil has been indicated mainly for three genera of microorganisms, viz., *Azotobacter, Beijerinckia,* and *Clostridium,* and (d) blue-green algae, Cyanophyceae, constitutes the second greatest possibility after legumes of supplying nitrogen to the soil through natural methods. Fixation of atmospheric nitrogen in rice culture by *Anabaena* and *Nostoc* has been studied; however, little research has been conducted on nitrogen fixation of terrestrial blue-green algae under conditions of dryland agriculture in tropical regions.

Considerable research is necessary to determine the importance of these other methods of nitrogen fixation and their adaptability to agricultural practices.

F. Mycorrhiza in Tropical Ecosystems

Mycorrhiza, or fungus-root, refers to a symbiotic association between some types of fungi and roots of higher plants. Most of the work on nutrient uptake by mycorrhiza has been conducted with ectomycorrhiza, in which the fungus forms a net around the root. The fungus component of mycorrhiza is thought to utilize sugars and other growth materials, such as amino acids and B vitamins from plant roots; in return it passes phosphate and other minerals to the roots.

The high efficiency of mycorrhiza in nutrient absorption is believed to be due mainly to increased absorbing surface from enhanced root proliferation, the presence of large absorbing surfaces on the fungal hyphae, and the greater rate of ionic absorption of the fungal sheath compared with the host root. Mycorrhiza are efficient in phosphorus absorption from the soil although it appears that the absorbed phosphorous is passed to the host in appreciable amounts only when the external supply of phosphorus is limited. In addition to phosphorus, mycorrhiza has been shown to be able to absorb metallic cations such as potassium and calcium. These indications of high efficiency of nutrient uptake by mycorrhiza are consistent with field observations. For example, the nutrition of some trees is known to be intimately tied to mycorrhizal development. For approximately 20 years in Puerto Rico, attempts to establish 26 pine species and strains on the island were completely unsuccessful until inoculum containing mycorrhiza from a pine stand in southeastern United States was introduced. The subject of nitrogen fixation by mycorrhiza is still incompletely known.

Soil conditions that favor the formation of mycorrhiza are found in many tropical soils, viz., adequate supply of organic matter, good soil aeration, and restricted supply of soil nutrients, particularly phosphorus. An intense photosynthetic rate by the host plant also favors mycorrhizal development. Went and Stark (1968) indicated that tropical rain forests have mycorrhiza, which decompose woody fruits and other litter on the forest floor, and also absorb and store soluble minerals which otherwise would be lost by leaching. Thus, the role of mycorrhiza in the soil-vegetation nutrient cycling may be of considerable importance in tropical forests.

V. Utilization of Forests

A major problem of tropical forestry at present is the widespread clearing of natural forests. Forests that have evolved over millions of years are disappearing in a matter of a few decades. These forests are rightly considered, at least over large areas, as a nonrenewable natural resource (Gómez-Pompa et al., 1972). With few exceptions, foresters and conservationists have been unable to stop this loss.

The disappearance of forests constitutes an irreparable loss of the very tool that is fundamental to many fields associated with tropical ecology. These communities are a source of many products and values unconnected with timber production. Natural communities constitute protective environments and are the sources for drug and medicinal plants. Loss of forests may not only eliminate or alter economic options, but also will affect wildlife, watersheds, airsheds, gene pools, and potential recreation land, all of which will increase in value as land becomes more intensively used.

The overwhelming nature of the problem of loss of forests calls for a concentration of research into those areas that are liable to reverse or stop this trend or, at least, which will justify the preservation of large and representative samples of the natural forest. Such sample reserves would be held inviolate until a better understanding of their real present and potential values becomes a fundamental consideration of decision makers and those entrusted to manage forest lands. The following topics are therefore suggested as meeting this fundamental objective.

Tropical foresters should assess the factors related to the evolution of sound forest- and land-management practices on national, regional, or biospheric levels. In particular, the location, class, and condition of natural forests must be inventoried, since we do not clearly know the actual extent of remaining forests, except in a very general way (Man and the Biosphere, 1972). These inventories should be made in the context of a sound land-use classification scheme. Further, man-made plantations should be compared to natural forest to determine whether they maintain soil and water quality. The nature and degree of human economic pressure on forest land, in particular those areas classified as not physically suitable for sustained agriculture or animal husbandry (presently called "marginal lands") should be determined. Finally, conservation attitudes and public awareness of natural forests and their potential in promoting development should be assessed.

These assessments involve the analysis of numerous biological and cultural variables and their interactions and can best be dealt with by case studies leading to predictive models. The basic ecological processes which underlie forest management have been discussed in the earlier sections of the report; here we concentrate on the man-forest interactions.

It is believed that much of what is happening in tropical American forests today depends not so much on the directives given by professional foresters, planners, government officials, or even the legal owners, but rather by the people, who are poor and uneducated and who live in or close to the edge of forests. Their motivation, actions, and other influences on the forests are poorly known and are often misinterpreted. Sociological and anthropological research is urgently needed to elucidate this important variable. This type of information is especially needed to improve land-use classification, so that the role of forestry can be adequately integrated with local customs and beliefs.

Wasteful utilization schemes and the pernicious influence of the "cut and get out" enterprises are at the root of the continuous encroachments on natural forest. These processes lead not only to gradual destruction of the resources both directly and indirectly by opening access roads to permit clearing for agriculture or grazing but are very often responsible for high timber prices and periods of acute scarcity. We must determine the kind of enterprise most able to practice sustained yield in specific areas. Research is urgently needed on the advantages and disadvantages of large companies with large capital investment, in comparison with small enterprises with limited capital, regarding such factors as sustained productivity, influence on nearby natural forests reserved for noncommercial timber production, and quality of life of those dependent on the industry.

Protecting tropical forests through vigorously applied land-use schemes, relieving pressures through various forest-plantation or forest-improvement devices, and forest legislation are all procedures to attain the goal of protecting large representative sample of forest. Improving timber exploitation by sound sustained-yield management practices may help but will possibly not save natural forests from further destruction. Wider utilization and better marketing of the many tropical woods that presently have little commercial value will also prove to be a very rewarding research topic. It will help to establish that management for wood production can successfully compete with alternative land uses.

Research is urgently needed to find new ways to produce food and other products in the wet tropics. Through intensive appraisal of current practices of utilizing leaves (including shoots), fruits, and flowers as well as other products that can be commercially harvested (medicinal products, gums, resins, ornamentals) from natural communities, new sources may be developed. Moreover research to convert the large productivity derived from leaves and wood into edible products (for man or domestic animals) through physical-chemical processes deserves to receive much more attention (Gore and Joshi, 1972). Although some of these processes may not be economical at this particular time it is still necessary to understand the limitations of the use of forest products, especially if conventional food and production does not keep pace with population increases. Obviously, these industrial conversion schemes may be advantageously combined with other industrial complexes that utilize tropical-forest products, particularly pulp and paper mills and large timber-processing plants.

Forest management for recreation, wildlife, scientific and educational tourism and for preservation of valuable species, gene pools, and various protective purposes (viz., water, soil, wildlife, etc.) has presently received little attention. Whereas the tools for this type of management are relatively well known, foresters trained in or oriented toward timber production are often unable to manage for conservation purposes. Analytical case studies of alternative land-use schemes are urgently needed.

VI. Utilization of Native Animals

Man interacts with a host of native animals in the American tropics in a variety of complex cultural, economic, ecologic, and epidemiologic contexts. Nevertheless, one is struck by the lack of scientific inquiry into this important human ecological phenomenon. The limited research is sparsely reported in the geographical and anthropological literature and to a lesser degree in ecological journals. The principle reviews of the status of knowledge have been made by Bennett (1967, 1971), Sternberg (1968), Gilmore (1950), and Harris (1965). These papers provide an insight into current and past literature on the subject and comprise a reasonably comprehensive statement regarding the present state of knowledge.

The following interactions between man and native animals are considered important areas for discussion and further research.

A. Subsistence Hunting and Fishing

Native animals have traditionally played a major role in the diets of subsistence peoples in tropical America and still do among a limited number of Amerind groups (summarized by Nietschmann, 1971). Nevertheless, few quantitative data are available regarding the principle taxa taken, their respective nutritional values, and the hunting and fishing methods employed. This general statement is true for contemporary societies as well as for earlier groups dating back into the pre-Columbian period. Bennett (1962, 1968), Coe and Flannery

(1967), have provided data on the role of animals in pre- and post-Columbian Amerind diets in Panama, southern Guatemala, and coastal Nicaragua, respectively. But the data for the general area of the American tropics are clearly insufficient for documenting with any degree of accuracy the historical trends and the regional variations concerning the ecological impact of subsistence hunting on native animal populations. Nevertheless it is clear that peasant people have the capacity to reduce significantly native animal populations. The general trend is clear, viz., there is a decreasing level of human consumption of animal protein derived from native animals as these are being greatly reduced in numbers, some to the point of being critically endangered (see Bennett, 1962; Daugherty, 1972; Doughty and Myers, 1971). This loss of high-quality protein has not been replaced by protein derived from domestic livestock, and as a result protein malnutrition is increasing to the point that some peoples of the American tropics are among the most malnourished in the world.

The possibility of management of native animals is being pursued (see Ojasti, 1971; Ojasti and Medina-P., 1972). However little is known of the sustained yield harvesting rates as a function of land use, season of cropping, age structure, and sex ratio of the crop. Because of the intrinsic taxonomic diversity it will undoubtedly be important to assess the optimal level of diversity relative to various levels of yield. The possibility of domesticating certain species and using others as quasidomesticates to diversify and improve the current meat and skin enterprises requires investigation.

B. Commercial Hunting and Fishing

Market hunting varies widely in its importance from one part of tropical America to another and national or local laws governing these activities are also varied. However, there are few reliable published data on the extent of market hunting for many tropical American countries and its impact on the fauna. The only way such data may be obtained is to conduct well-planned field investigations in selected areas where hunting for meat is restricted and compare these to areas of exploited hunting. However, until *in situ* studies are made, legislation restricting such activities will be slow in enactment.

Of all aspects of commercial exploitation of wild animals in tropical America, fishing provides the greatest quantity of numerical data but this must not obscure the fact that there are major gaps in our understanding of this important activity. Major saltwater fisheries are frequently the object of close governmental scrutiny for fiscal purposes and are thus productive of fairly accurate data as with the case of the Mexican shrimp fishery and the anchovy fishery in Peru. Economically minor saltwater fisheries, however, are frequently not reported upon in any useful detail. Such fisheries, however, demand study because of their ecological importance.

Commercial freshwater fishing has been largely ignored as an object of study even though it is significant in much of tropical America. The major area for this activity is the Amazon basin, and here the single most important fish appears to be the pirarucu *(Arapaima gigas)*. Although referred to in many publications (for example, Couto de Magalhaes, 1931), we have been unable to locate a published general ecological study of the pirarucu or other freshwater commercial fisheries in the Amazon area.

Another problem which deserves close monitoring and research is the commercial export of tropical wildlife, particularly from the Amazon basin for zoological gardens, pet shops, research centers, and to satisfy commercial demands for exotic furs and skins. Such commercial exploitation is currently exerting inexorable demands upon Amazonian wildlife (Doughty and Myers, 1971).

C. Human Alteration of Animal Habitats

The alteration of animal habitats by human activity is clearly the major cause of the depletion of numerous animal populations in the American tropics. At the present time there are specific environments that are clearly threatened and even endangered as a result of human pressures. These particularly include the intertidal mangrove forests and the upland cloud forests, both of which are spatially restricted environments and therefore particularly susceptible to human disturbance (Daugherty, 1973a). The general aspects of mangrove- and cloud-forest ecology are known, but better documentation of the ecologic, economic, cultural, and scientific values of these ecosystems is needed.

Based on present knowledge, some general statements can be made on the ecological effects of habitat alteration, and on the observed decline of native animals but impact in the context of the whole system is poorly understood. Less obvious repercussions that deserve major research attention include alterations of predator-prey relationships, disruptions of food webs and other community relationships, and declining diversity (Daugherty, 1973b). Recent evidence from New Guinea (Diamond, 1973) and Barro Colorado Island in the Canal Zone (Terborgh, 1973) indicate that the survival of numerous tropical birds, and perhaps other animal groups as well, are dependent on the maintenance of large areas of continuous rain forest due to their low reproductive capacity and limited ability for dispersal across open areas. Thus, a major research need is to define species-area relationships for tropical animals not only in rain forests, but in other terrestrial ecosystems.

D. Wildlife for Tourism and Sport Hunting

The most striking zoologic characteristics of the neotropics are the extreme diversity of species and the large number of endemic taxa. The diversity and unusual nature of the neotropical fauna including the avifauna, coral reefs, and freshwater fisheries constitute a potentially valuable wildlife resource for tourism and sport hunting. Leopold (1959), for example, has shown the economic value of sport hunting in Mexico. Research is needed to assess the potential for developing tourism and sport hunting in other areas of the American tropics and to assess the environmental hazards associated with such activities. The economic success of the wildlife-based tourist industry of East Africa is well known. Kenya, for example, currently derives a major share of her foreign exchange earnings from tourism. But, unlike the East African savannas, the neotropical grasslands are notably lacking the diverse ungulate fauna upon which much of the African tourism is based. Furthermore, many of the neotropical species that are attractive for tourist and sport purposes are clearly endangered. This is particularly true for the American felids, five species of which are listed on the Convention on International Trade in Endangered Species of Wild Fauna and Flora (International Union for Conservation of Nature and Natural Resources, 1973).

In El Salvador alone, sport hunting has caused the extermination of three species of cats as well as the critical depletion of numerous other economically valuable game animals, such as the white-tailed deer, caiman, and crocodile (Daugherty, 1972).

Undoubtedly the American tropics offer some potential for increasing sport hunting and tourism based on wildlife, but it is clear that the development of a wildlife-tourist industry in the American tropics must be undertaken with extreme caution and may not be patterned after the African experience. Currently lacking are sufficient data on American tropical animals regarding the degree of hunting and tourist pressures that they can withstand. Also the long-term security of wildlife resource may depend on its acceptance as a value item in the social mores of the local population (Parker and Graham, 1971; Harris and Marks, in press).

Several international symposia have dealt with the conservation and utilization of tropical wildlife resources, including those of the neotropics. Those of particular importance were the International Conference on the Utilization of Wildlife in Developing Countries (Bad Godesberg, West Germany, 1964), the UNESCO Conference on the Use and Conservation of the Bisophere (Paris, 1968), the Simposio y Foro de Biología Tropical Amazónica (Florencia and Leticia, Colombia, 1969), and the Simposio sobre a Biota Amazonica (Belem, Brasil, 1966). A consistent theme of these symposia was that there is insufficient ecologic data available on many tropical animals to provide a sound framework for rational wildlife management for sport and tourism purposes. There is also a lack of understanding of the cultural and esthetic values that various societies in the American tropics place on their wildlife resources.

It seems clear that any sound and lasting resource management policy is dependent upon the close interaction of three major factors: (a) the physical and ecological basis of the resource, (b) economic considerations as well as nonmonetary value systems, and (c) the social and institutional structures that ulitmately dictate the acceptance or rejection of otherwise sound policy. Unfortunately, the fundamental descriptive knowledge in most of these areas is lacking, and there is a need for basic ecological information in order to project which types of utilization schemes are feasible.

VII. Selected Industrial and Developmental Activities

We have not treated industrial or urban concentrations in this report because of limitations in time and available funding for study of such an extensive subject. However, hydrological projects, transportation, mining, and tourism have direct impact on rural and natural environments and are considered.

A. Hydrological Projects

Development in agriculture and industry requires control of water supplies and power derived from hydroelectric projects. Ten percent of the world's stream flow is currently regulated (United Nations, 1970), and one recent estimate (Szestay, 1972) states that by the year 2000 approximately two-thirds of the world's stream flow will be regulated. Due to accelerating demands for hydropower, irrigated agriculture, flood control, aquaculture and lake fisheries, river impoundment, and irrigation projects rapidly are being extended to many still free-flowing river basins in the tropics. Historically, the majority of these projects have been concentrated in Africa. Currently they are receiving increasing attention in the Asian and American tropics.

In addition to the increasing demands for hydropower, growing populations require more food, raw materials, and energy resources. River-basin development is receiving increasing attention as a means for satisfying these accelerating requirements. Initially, many hydrological projects were conceived as single-purpose projects, usually for power or irrigation water. More recently, the size and scale of these developments have expanded, and projects have sought to provide multiple benefits. The Aswan, Volta, Kariba, La Plata, Papaloapan, and Mekong projects are all examples of this integrated development approach, often involving multiple dams, irrigation works, and related structures.

Unfortunately, few of these major development projects include careful scientific study of their ecological consequences. Until thorough scientific investigations of social and environmental costs and benefits in river basin development are included as an integral part of development planning, unforeseen and costly perturbations are likely. In some projects, evidence points to a number of such costly secondary effects of sufficient magnitude to lead one to question the wisdom of initiating the development (Favar and Milton, 1972). In other cases, ecological research could have provided guidelines to preclude the development of serious social and environmental disruptions and costs. Since river-basin development projects are proliferating faster than scientific investigations can predict their total environmental impacts, the need for accelerated ecological research on such projects is imperative.

Numerous negative impacts have now been identified through research and case studies of many of the early African and several Asian and tropical American river impoundments. Negative modifications of tropical river basins have most notably involved resettlement of human populations in upland areas, aggravation of water-borne diseases, disruption of riverine fisheries, spread of aquatic weeds, salinization and/or water-logging of irrigated lands, siltation of reservoirs due to inadequate controls of watershed land uses, inundation of important natural, historical, and archeological sites, and triggering of earthquakes in certain highly earthquake-prone reservoir sites (Dasmann, Milton, and Freeman, 1973).

Since most tropical American river basins have human communities that developed social systems of resource use based upon the productivity of natural river systems, it is imperative that ecological research and monitoring be applied through integrated planning, anticipating environmental problems and modifying development criteria, preinvestment surveys, and project planning and construction. In some cases, such research will involve modifications of project siting and design. In others, alternative development strategies may be advisable. The following research priorities are suggested to help identify critical environmental problems where ecological research can help in modification of tropical river-basin projects.

1. Fishery research

Reservoir development in tropical river basins may provide a new lake environment for a fishery. The impacts on upstream and downstream fisheries, however, are commonly deterimental. Dams act as barriers to fish movement and migration to feeding and spawning areas. They also change riverine nutrient loads, temperature, natural flooding cycles, sediment levels, and dissolved oxygen content. Another common problem is reduction of nutrients for coastal marine fisheries dependent upon sustained nutrient inflow from rivers.

Development of reservoir fisheries, however, are beset with a different series of ecological limitations. Most lake fisheries must initially develop from a natural stock of species adapted to riverine, not lacustrine, conditions. Therefore, careful ecological study of projected local reservoir conditions and species adapted to them is necessary to optimize production and utilize the many new riches created by the reservoir environment. Introduction of exotic species may be advisable in some cases, but such introductions can also cause severe problems of infestation. Utilization of native species should therefore receive first research attention, and if exotics are contemplated, the possible secondary impacts of introduction must be carefully studied.

An area of high research priority is the study of fish ecology, migration, and potential impact from reservoir projects in river basins likely to be modified. Where reservoir development is imminent or planned, research is needed on the dynamics of nutrient supply, temperature, and other factors in maintaining productivity of estuarine and marine ecosystems. Research on the role of riverine fish populations in the diet of river-basin cultures, methods of catch, and possible cultural impacts from decline of riverine production from reservoirs also should receive high priority in streams likely to be modified. Devices (such as fish passes) that could minimize adverse effects also need attention in relation to life histories of critical species.

Of particular importance to river-basin development are long-term studies on the potential for aquaculture. The potentials for increasing yields from fish, shellfish, aquatic plants, turtles and crocodilians, and waterfowl by careful manipulation of aquatic ecosystems are substantial and deserve priority attention in future development planning for tropical river basins. Yields from fish ponds, for example, may exceed both the economic and nutritional yields possible from the same site in agricultural production (Lagler, 1969). In addition to fish ponds, mixed irrigated agriculture and fish systems, duck production in lakeside ponds replenished by raising lake levels, and utilization of aquatic plants for food and fertilizer all deserve ecological investigation.

2. Water-borne disease

Reservoir construction and stabilization of irrigation schemes have, throughout the tropics, tended to spread and intensify problems of water-borne human and domestic animal disease. In addition, many of the transportation systems, resettlement programs, and nutritional changes associated with such projects have adversely affected the pattern of disease. Recently a number of case studies of tropical reservoir projects have clearly portrayed the relationship between water-borne disease and riverine projects (Farvar and Milton, 1972). These studies re-emphasize the vital necessity for all tropical river-basin projects involving water stabilization to require intensive ecological studies early in the planning process.

One of the most persistent water-borne disease problems is schistosomiasis (bilharziasis), which in the American Tropics is found from Brazil to Venezuela and in parts of the Caribbean Islands. *Schistosoma mansoni* is the only species currently responsible for severe problems in the Americas. It was apparently introduced by the African slave trade and European military contact after campaigns in Africa. Schistosomiasis is transmitted by an intermediate aquatic snail host carrying the miracidia, which develop into free-swimming cercariae that leave the snail and infect the human host by direct penetration of the skin or through drinking. At this point the schistosomes migrate throughout the human body, causing damage to different organs and eventually (in the case of *S. mansoni)* move to the human intestine where they pass their eggs out with the feces to later hatch and re-infect snail populations.

Recent increases in density of rural human populations, poor sanitation, and the stabilization of irrigation and reservoir impoundments (which allow intermediate host snails to spread) have led to a spectacular worldwide rise in schistosomiasis. If the spread of this debilitating disease is to be controlled, research on the ecology of this and other water-borne diseases must receive high priority. Particular regional emphasis is required where potentials exist for dramatic increases from existing foci, as in the colonization of the Amazon Basin by settlers from infested northeast Brazil and where reservoir and irrigation projects are planned in regions known to harbor the disease and suitable intermediate snail hosts. Research on the suitability of native American snails as potential host species for schistosomes, on aquatic limiting factors to host snail distribution and density, and on preventive techniques for spread of suitable snail hosts (such as biological controls, water level manipulations, chemical control and its side effects) are also of high priority.

Similar research is also needed for other forms of diseases borne by water at some stage in their life cycles, as discussed earlier in Section 2. Reservoirs and irrigation works, for example, may serve as breeding grounds for malarial mosquitos and a variety of mosquito-borne viral infections. Research on the distribution of such diseases should be correlated with studies of potential impacts on the disease environment from water impoundment and diversion projects.

3. Aquatic weeds

The explosive growth of aquatic weeds is one of the most common impacts of tropical reservoir and stabilized irrigation works. Water hyacinth *(Eichhornia crassipes)* and water lettuce *(Pistia stratiotes)* are two free-floating vascular plants commonly implicated. Losses from such aquatic plant growth usually come from one or a combination of the following processes: (a) competition with fish for nutrients and light sustaining fish-food sources (such as algae), (b) physical interference with fishing, (c) disease impacts by providing habitat for disease-bearing snails and mosquitoes, (d) evapotranspiration losses, and (e) interference with lake recreation and navigation.

Controls for aquatic weeds are still poorly developed. Herbicides have been utilized for control in some lake systems, but only at a high cost. In Surinam's Lake Brokopondo, for example, filling occurred in 1964. By 1965, 17,900 acres were covered by water hyacinth. By April 1966, 41,200 acres were covered (53 percent of the total lake surface). At this point a 2,4-D control program was initiated at an annual cost of US $250,000 (Dasmann, Milton, and Freeman, 1973). Other feasible control techniques include removal by hand and machine and by boat and draglines, shoreline drawdown, disease organisms and insect, fish mammalian, and snail pests. Much more research on biological and mechanical controls are needed, however. No inexpensive mechanical control is yet widely applicable and little is known about the ecology and behavior of pests, diseases, and herbivores that might control aquatic weeds.

One promising approach is utilization of aquatic weeds as a resource. Some programs, particularly in Asia, have used aquatic plants for removal of organic wastes from polluted water, as a crop providing fish, cattle, pig, buffalo, and poultry food, for mechanical extraction of leaf protein for human food, and as a source of fertilizer and industrial raw material. The value of this approach is that it views aquatic plants as a positive resource, which could turn an unwanted cost into a valuable economic asset in many areas.

A primary avenue for research on aquatic weeds is in the careful study of the distribution and life cycles of aquatic plants likely to invade man-made lakes and irrigation works in the American tropics. Such research should include (a) study of natural and introduced disease, pest, and cropping organisms, (b) impacts of aquatic weed growth on fisheries and other freshwater organisms, (c) physical interference with fishing, navigation, and recreation, (d) water-borne disease relationships, and (e) effects on evapotranspiration rates.

4. Watershed management

Riverine dams and irrigation systems have an intimate relationship to their watersheds and to the supply of fresh water they provide.

As agents of erosion, rivers and streams are extremely sensitive to alterations in watershed land use that change rates and timing of streamflow, sediment load, deposition, and related factors, as mentioned in Section 3. Over time, biological communities have adapted to local ecological systems in ways that have tended to reduce rates of erosion. Many traditional land-use patterns in the tropics, such as low-density shifting agriculture, terraced agriculture, and mixed tree-shrub-ground cropping systems have also tended to minimize erosion and runoff difficulties.

Today, however, with rising human densities on rural lands, or rapid increase in road-building, the exposure of many soils to erosion from spaced plantation culture, and settlement of formerly forested marginal hill land, rates of erosion in tropical America have increased greatly. Increased soil erosion has had and continues to have serious impacts on soil productivity. It has also increased the rates of runoff leading to accelerating problems of river flooding in many areas.

At the same time, however, man's increasing dependence on riverine reservoirs and irrigation systems for high-production agriculture, hydropower, fish protein, and other benefits poses new needs for adequate controls on watershed erosion. Recent studies in Colombia of the Anchicayá Hydroelectric Project (Allen, 1972) document the relationship between destruction of the watershed from uncontrolled settlement spurred by the new transportation systems giving access to the reservoir area and rapid siltation of the reservoir (also see Fig. 3.4).

The requirements for watershed management vary according to climate, soils, vegetation, and land-use patterns. Arid watersheds, for example, commonly are utilized for grazing and for sources of wood for household fuel. Unless grazing and cutting are managed to retain sufficient soil cover, erosion and flooding during rainy periods are likely. Similar erosion problems are initiated on subhumid and humid watersheds where steep slopes are cleared for agriculture or timber, or where agricultural runoff controls are ignored.

These problems have been summarized as follows (Dasmann, Milton, and Freeman, 1973, p. 210):

> Any change in the watershed area will affect the flow of water and the quantity of sediment carried downstream. The protective role of forests in maintaining a relatively stable flow of water for reservoir impoundments and irrigation works is well-established. Not only does protective vegetation decrease flood peaks, but it also usually increases water discharge during dry periods. Forests and their root systems are also of great importance in preventing erosion and reducing sediment loads in streams and rivers.

Major gaps in research currently exist in tropical watershed management improvement. Research is needed on the interrelationship between rainfall, soils, vegetation, and alternate land-use patterns in major tropical river basins undergoing development to determine optimal land use methods and management. Particular emphasis is needed to coordinate the multiple goals of sustained yields from agriculture, grazing, or timber with the goals of flood prevention, erosion control, and prevention of lake sedimentation.

Often the goals of flood prevention, sustained water flows, and erosion prevention are highly compatible with regional needs for preservation of areas for wildlife and natural vegetation, park-based tourism, protection of natural gene pools, and other conservation values. Integrated research to determine the combined economic value of protecting watersheds for such purposes is needed. Further research on planning and management techniques to optimize these multiple goals is of high priority.

Research on the impact of lake sedimentation on loss of storage capacity, reservoir life, lake recreation, fishing, and other lake resources is needed. Such analysis should include translation into economic data of the combined economic costs of sedimentation through specific case studies. Further problems may come from downstream alteration of water regimes, loss of soil nutrients from seasonal flooding, channel and delta erosion, and increased waterborne disease (Scudder, 1972).

The downstream impacts of reduced silt loads and nutrient flows trapped by reservoirs needs research. Studies should include: impacts on downstream freshwater fisheries and estuarine and marine fisheries, effects on downstream agriculture, and the impact on undermining bridge abutments, channel erosion, and geomorphological equilibria in delta zones.

Where direct losses of forest due to the inundation by man-made lakes are probable, research on their ecology and economic value should be part of the prefeasibility research of the project.

5. Recreation and tourism

Construction of man-made lakes in the American tropics present opportunities for recreation and tourism. Properly designed and managed, such reservoirs provide swimming, boating, fishing, hunting, wildlife observation, camping, and nature education. To maximize these multiple benefits, the primary objective of the reservoir may have to be modified to provide better waterfowl nesting and fishing conditions, to prevent water-borne disease, to promote sport-fishing species production, and to consider other such factors relevant to multiple recreation goals. Creation of lake-edge wildlife and forest-watershed nature reserves also can play an important role in attracting recreation and tourism.

However, reservoirs often inundate areas of high touristic and recreational value as parks and reserves, scenic free-flowing streams, canyons, waterfalls, rapids, and important plant and animal communities. Similarly, areas of high archeological or historic significance may be inundated. Lastly, biotic resources of native fish, wildlife, and plants used as food and raw material by local human communities may be lost. As mentioned earlier, the need to survey and estimate the value of these environmental resources before the project is initiated is critical.

High priority must be given to the survey and inventory of natural areas, scenic sites, and biotic, scientific, or historic resources in developing tropical river basins. Where such resources would be lost by inundation or related activities (such as road-building), the costs of their loss should be determined and included in development feasibility analysis. Costs of salvage, study, and restoration of these resources ought also to be included in development planning.

6. Human resettlement problems

Among the most difficult and costly problem areas related to tropical man-made lake development is the inundation and resettlement of human populations and communities. Many projects have flooded out substantial populations and inundated their arable land. Following resettlement these people commonly face a number of new problems: break-up of families and social units; resettlement on upland sites with poor soils, limited agricultural potential, loss of former fish-protein sources, inadequate potable water, poor transport, exposure to new forms of disease, and malnurition; lack of funds, planning expertise, and personnel for planning and managing the resettlement; inadequate education and training programs to help people adapt to the new site; inadequate compensation for losses incurred; land competition with previous settlers of the resettlement site; unplanned spontaneous colonization of critical watershed areas causing damage to reservoir values; lack of research on new fishery, crop, and livestock opportunities both ecologically and economically feasible; lack of medical assistance; little opportunity for local community input to planning and managing the resettlement operation (Dasman, Milton, and Freeman, 1973).

The size and scale of many tropical reservoirs and their numerous resettlement problems make ecologically sound and culturally equitable planning an extremely complex issue. For these reasons, ecological and social research to help anticipate and solve these problems should be initiated at the earliest possible stages, even at the feasibility study stage. In some cases, the fair estimation of ecological and social costs may mitigate against reservoir development and suggest alternative patterns of development with fewer social costs, high benefits, and better distribution of both costs and benefits. The role of ecological research, both in obtaining full estimates of social costs and benefits and in helping derived alternative development patterns, is absolutely essential.

In relation to research in this area particular attention is needed on (a) soil capabilities, climate, and native biota, (b) alternative patterns of feasible garden and market crop production, (c) local epidemiological problems, (d) methods to diversify subsistence crops, perhaps mixed with domestic animal and fish production, to provide nutritionally balanced diets, (e) village planning and house design based upon traditional cultural requirements, (f) sources of potable water for household use and irrigation, and (g) possible land-use conflicts with already established groups.

7. Irrigation projects

Salinization and soil drainage problems are a particular problem in arid irrigated areas, such as the Imperial Valley in California and Pakistan's Punjab. Subsequent invasion by salt-concentrating plants is a closely related problem. Flushing with water, regulation of irrigation flow, and pumping to lower the water table are common techniques to combat the problem. Applied ecological research is needed to identify areas prone to salinization early in irrigation planning, to estimate potential for salt-concentrating plant invasion, and to determine the effects of salt contamination of downstream freshwater on other agricultural areas and aquatic life. Ecological research is also necessary for irrigated areas where water contamination by pesticides, herbicides, domestic sewage, and water-borne disease is likely to occur.

B. Transportation

Transportation is a key process in the development of industry, modern agriculture, forestry, and tourism. Yet, many large and small regions of tropical America possess almost no means of regular access by surface or by air. Many other areas are inadequately served by trails, dry-season roads, river transport, or very small air-strips. Even the most cursory examination of road maps of almost any part of tropical America reveals that developed transport route nets are conspicuous by their absence. And in many of the areas where all-season transport routes exist they are of limited utility and confined to a single all-season road or airstrip. Such realities have exacerbated the formidable problems associated with improvements in agriculture and animal husbandry and the amelioration of undesirable social conditions. Because of these problems, increasing attention is being given to the funding of road construction throughout the American tropics. The most notable examples are the Trans-Amazonian Highway currently under construction, the partially completed Carretera Marginal from Venezuela to Bolivia in the western Amazon, and the completed Brasilia-Belem and Brasilia-Acre Highways.

Experience has repeatedly shown that a variety of ecological changes occur when roads are constructed into or through areas previously lacking them. The most significant alteration is the replacement of rainforest by ecosystems which can, if improperly managed, result in accelerated runoff and erosion, a decline in soil fertility, a change in soil structure, the loss of actual or potential raw materials, and the destruction of economically valuable species of wildlife. The general trend along penetration roads is initial clearance for shifting agriculture, subsequent abandonment of farmland after a few years of cultivation, and finally the consolidation of the original small holdings into small- to medium-sized ranches which maintain the vegetation cover in the form of open to dense scrub savanna (Denevan, 1973).

Up to and including the present it has been almost entirely a matter of *post hoc* recognition that a new road was responsible for initiating direct and/or indirect ecological changes in the region through which it was constructed. Because it is to be expected that road construction will continue to receive a high priority in development planning, it is obvious that research is needed relative to elucidate the ecological phenomena associated with road building. And while road building dominates the development of transport routes, similar attention should be given to other major aspects of transport development such as air fields and water transport.

C. Mining

Mining is an important industry in many areas of tropical America. Long before the first Europeans arrived native peoples of tropical America engaged in mining activities. Although modest by contemporary measures, these included drift mining for native copper in the Andes and the extraction of gold from alluvial deposits. The environmental influences of such activities may have been minor but there has been little research on which to base a conclusive evaluation.

With the arrival of the Europeans a new chapter in mining began, marked by the introduction of new techniques and a larger range of sought-after minerals and metals. But gold, continued to be the most highly prized mineral, just as in pre-Columbian times. On the other hand, in terms of the environmental implications as well as the monetary return, silver emerged as the major element once the preliminary colonial mineral exploration was completed. The major areas of large-scale colonial mining were located in Mexico and in the Central Andes.

What ecological modifications were associated with the mining and smelting of silver in the Andes (or the other locations where they occur within the American tropics)? The fundamental modification seems to have been in the vegetation cover and especially with those tree and shrub species suitable for mine timbers and, more importantly, useful for charcoal manufacture, which in turn was often a part of the ore reduction process. Although it is logical to assume that the great silver-mining center of Potosi, Mexico, made enormous demands from a large region to supply it with the wood needed for fuel and mine timber there are no existing studies to corroborate this probability. Thus, the current and long-existing condition of treeless slopes in some silver-mining regions is often "explained" by invoking climatic limitations and the like. The cultural inputs in this and other silver-mining regions of the past need investigation in order to arrive at some evaluation of the relative roles of cultural, biological, and physical parameters in producing the vegetation cover of the present day.

Although colonial mining was focused on a limited array of precious metals and minerals (e.g., diamonds, emeralds), in more recent decades the emphasis has shifted to another set of metals and minerals and the scale of mining has increased. Since a complete listing of all the metals and minerals now mined in the American tropics would be tiresome, only some of those whose present scale is of a large magnitude will be mentioned. These include iron, copper, petroleum, bauxite, coal, and limestone. It might be noted in passing that the "royal pair" of the colonial period (gold and silver) rank well down the list today.

Mining techniques may be conveniently divided into the following categories: strip mining, open pit mining, alluvial or placer mining, hardrock or drift mining, and wells. Each of these techniques carries with it various implications respecting the kinds of environmental modifications that are likely to occur. Most of what is known at the present time about such modifications is derived from experience in the middle and high latitudes and it is important to know if mining techniques in the tropics leads to similar effects on basic environmental parameters. For example, bauxite mining is almost always done as a combination of open pit and strip mining. This procedure involves the removal of an overburden layer of varying depth and varying soil quality, with the consequent removal of the vegetation cover. Thus far, major bauxite mining activity has been more or less limited to Jamaica and the Guianas but the efforts presently given to bauxite exploration are clearly indicative that an extension of this mining may be anticipated, the recent low world aluminum prices notwithstanding. It is important to know if the Jamaican and Guianan experiences are sufficient to understand the ecological aspects of bauxite mining or if more research is required.

Iron ore of commercial quality is known to exist in only a few areas in the American tropics, but some of the more impressively rich and extensive iron-ore bodies in the western hemisphere are known in this region and are in part the object of exploitation. The ecological relations of open pit iron mining in tropical ecosystems is little known or understood at present.

Limestone is widely distributed in parts of the region and is increasingly exploited (when it occurs in economically practicable locations) for the production of cement. This mining essentially involves quarrying techniques. Although many such operations are of modest size and highly localized, others are of considerable magnitude. As they expand, they leave ever-larger depressions in which water may collect and in which some arthropod vectors of disease may find sites for completion of parts of their life cycles.

Coal and petroleum are mined at various locations in tropical America. Coal mining is of significance only in Colombia, where drift mining for anthracite coal is used, often at considerable depths where drifts follow narrow coal seams. The mining of petroleum is more diffusely distributed but major production is focused in a limited number of areas: the Venezuelan littoral including Lago Maracaibo, the newly developing field between Ecuador, Peru, and Colombia, the region near the Colombia-Venezuela border, and Trinidad. Continued exploration will undoubtedly lead to the discovery of new oil fields. Although much of current environmental concern concentrates on petroleum spillage in marine ecosystems, we must be sensitive to the fact that oil fields located away from marine environments open the way to spillage and other phenomena associated with oil production in ecosystems whose responses to such occurrences are not presently known. In addition, ancillary activities such as road building and pipeline routes represent situations where one must expect to encounter modifications in the ecosystems through which they are constructed.

From this brief discussion two elements respecting mining and environmental modifications have emerged: (a) that whereas mining tends to be localized, the intensity of mining and frequency of mine sites is on the increase (b) that we have at present mainly a generalized comprehension of the ecological relationships associated with mining in tropical America; however, it is clear that we will require the kinds of information described in Section 3 to judge the impacts of mining and other industry on the environment.

D. Tourism

Tourism also utilizes the natural environment and is of continuous, growing importance in tropical America. An important factor in this growth has been the desire to visit wildlife areas such as the Galapagos Islands, scenic areas such as Lake Atitlán and Igaçu Falls, areas of historical or archeological interest such as Tikal, Macchu Pichu and Cartagena, and areas offering diverse opportunities for outdoor recreation, such as coastal beaches, coral reefs, volcanoes and islands.

Increasing tourist demand for utilization of these natural assets has led to rapid development of many of these sites and improved road and air routes to hotels, swimming beaches, skin-diving, hiking, hunting, fishing and boating. Much of the tourism derives from international visitation and has provided significant new sources of foreign exchange for many countries in the American tropics. Mexico, for example, has estimated expenditures on tourism in 1980 will be at U.S. $1.6 billion (Dasmann, Milton, and Freeman, 1973, p. 113). Also in Kenya and other East African countries, for example, a large proportion of their national incomes derives from tourism attributed to their national parks, wildlife, regulated hunting, and outdoor photographic safaris. With proper planning tourism also is unique in its capacity to increase local job opportunities and incomes.

One of the major problems of modern tourism development has been a spectrum of activities that has led to various degrees of deterioration of the natural assets that were responsible for attracting tourists. Roads and hotels, for example, have often been located too close to areas of outstanding recreational value; access roads have often caused erosion and sedimentation in natural areas, spontaneous colonization of park sites, and deterioration of the visual environment; mass-recreational tourism in fragile zones has caused dune erosion, destruction of native vegetation from trampling, and the disappearance of local wildlife because of noise and other forms of disturbance; coastal beaches and reefs have been degraded by sewage, marinas, pollution from human settlements, siltation from erosion in nearby watersheds, dredging activities, and marine oil spills.

Most of these negative environmental impacts have led to serious disruption of the resource base upon which tourism development depends. In almost all cases, attention to the application of ecological research and principles in planning and development phases could have either prevented or greatly diminished the negative impacts.

The concern over park deterioration and the loss of those areas suitable for new parks underscores the pressing need for accelerated ecological research contributing to location of new park sites, improved park planning and management, setting of limits to human use of parks, and the zoning of human use areas within parks. Some of the kinds of information required for these needs are discussed in Section 2. Increasingly, intensive tourist development is seen as incompatible with the primary purposes of national park protection, and master recreational planning is attempting to divert mass recreational demand to less unusual and fragile areas. Special studies are needed on fragile, unique and endangered environments to determine national and international priorities for protection. Such research should investigate planning, management, and zoning needs based upon ecological and aesthetic factors determining usage.

Little research has been done on the variation of local cultural resilience to tourist impact. Some cultures, such as the Otovalo in Ecuador, are remarkably adaptive to tourist impact and have been able to retain much of their culture. Other cultures tend to be much less resilient to tourist impact. Research is needed on the mechanisms involved in such cultural resilience or fragility to allow better tourism planning in anticipation of impacts.

References

Allen, R. 1972. The Anchicayá hydroelectric project in Colombia: Design and sedimentation problems, p. 318-342. *In* M. T. Farvar and J. R. Milton (eds.) The careless technology: Ecology and international development. Natural History Press, Garden City, New York.

Barley, K. P. 1961. The abundance of earthworms in agricultural land and their possible significance in agriculture. Adv. Agron. 13: 249-266.

Bennett, C. F. 1962. The Bayano Cuna Indians, Panama: An ecological study of livelihood and diet. Ann. Assoc. Amer. Geogr. 52: 32-50.

Bennett, C. F. 1967. A review of ecological research in Middle America. Latin Amer. Res. Rev. 2: 3-27.

Bennett, C. F. 1968. Human influences on the zoogeography of Panama. Ibero-Amer. 51: 1-112.

Bennett, C. F. 1971. Animal geography in Latin America. p. 33-41. *In* B. Lentnik, R. L. Carmin, and T. L. Martinson (eds.) Geographic research in Latin America. Benchmark, 1970. Ball State University, Muncie, Indiana.

Borden, R. J. 1941. Cane growth studies. Factors which influence yields and composition of sugar cane. Hawaiian Planters Rec. 45: 241-263.

Bradfield, R. 1969. Intensive multiple cropping. Int. Rice Res. Inst., Coll. Agric., University of the Philippines, Los Banos. 2 p (mimeo).

Bradfield, R., and R. O. Miller. 1954. Soil structure. Trans. Fifth Int. Congr. Soil Sci., Leopoldville, Congo. 1: 131-145.

Chang, J. 1970. Potential photosynthesis and crop productivity. Ann. Assoc. Amer. Geogr. 60: 1-92.

Coe, M., and K. Flannery. 1967. Early cultures and human ecology in south coastal Guatemala. Smithsonian Contr. Anthropol. 3: 1-136.

Couto de Magalhães, A. 1931. Monografia brasileira de peixes fluviales. "Graphicars" Romiti, São Paulo.

Dasmann, R. F., J. P. Milton, and P. H. Freeman. 1973. Ecological principles for economic development. Wiley, New York.

Daugherty, H. E. 1971. Climate and ecology in Latin America, p. 47-71. *In* B. Lentnek, R. L. Carmin, and T. L. Martinson (eds.) Geographic research in Latin America-Benchmark 1970. Ball State University, Munice, Indiana.

Daugherty, H. E. 1972. The impact of man on the zoogeography of El Salvador. Biol. Conserv. 4: 273-278.

Daugherty, H. E. 1973a. The accelerating conflict between economic demands and regional ecological stability in coastal el Salvador p. 115-124. *In* A.D. Hill (ed.) Latin American Development Issues. CLAG Publ., East Lansing, Mich.

Daugherty, H. E. 1973b. The Monte Cristo cloud forest of El Salvador—a chance for protection. Biol. Conserv. 5: 227-230.

Daugherty, H. E. 1974. El hombre y el medio-ambiento en El Salvador. Una historia ecologica. Direccion de Publicaciones, Ministerio Educacion, San Salvador. 302 p, in press.

Deneven, W. M. 1970. Aboriginal drained field cultivation in the Americas. Science 169: 647-654.

Deneven, W. M. 1973. Development and the imminent demise of the Amazon rain forest. Prof. Geogr. 25: 130-135.

Diamond, J. M. 1973. Distributional ecology of New Guinea birds. Science 179: 759-769.

Doughty, R. W., and N. Myers. 1971. Notes on the Amazon wildlife trade. Brit. Conserv. 3: 293-297.

Ewel, J. 1971. Experiments in arresting succession with cutting and herbicides in four tropical environments. Unpublished Ph. D. thesis, University of North Carolina, Chapel Hill.

Farvar, M. T., and J. P. Milton. 1972. The careless technology: Ecology and international development. Natural History Press, Garden City, N. Y. 1030 p.

Gilmore, R. 1950. Fauna and ethnozoology of South America, p. 345-464. *In* J. H. Steward (ed.) Handbook of South American Indians. Vol. 6. Smithsonian Institute, Bureau of American Ethnology, Washington, D. C.

Gómez-Pompa, A., C. Vázquez-Yanes, and S. Guevara. 1972. The tropical rain forest: A nonrenewable resource. Science 177: 762-765.

Gore, S. B., and R. N. Joshi. 1972. The exploitation of weeds for leaf protein protection, p. 137 to 146. *In* P. M. Golley and F. B. Golley (compilers). Tropical ecology, with an emphasis on organic production. University of Georgia, Athens.

Greenland, D. J. 1971. Changes in the nitrogen status and physical condition of soils under pastures, special reference to the maintenance of the fertility of Australian soils used for growing wheat. Soils Fertilizers 34: 237-251.

Harris, D. R. 1965. Plants, animals, and man in the outer Leeward Islands, West Indies. Univ. Calif. Publ. Geogr. 18: 164.

Harris, L. D., and S. A. Marks. 1974. An appraisal of African game conservation. *In* M. T. Farvar (ed.). AAAS Symp. on Envir. and Int. Develop., in press.

Henzel, E. F., and D. O. Norris. 1962. Processes by which nitrogen is added to the soil/plant system. Common. Bur. Pastures Field Crops, England. Bull. No. 46. p. 1-18.

International Union for Conservation of Nature and Natural Resources (IUCN). 1973. Convention on international trade in endangered species of wild fauna and flora. Morges, Switzerland. 12 p.

Johannessen, C. L. 1963. Savannas of interior Honduras. Ibero-Amer. 46: 1-160.

Lagler, K. F. (ed.). 1969. Man-made lakes; planning and development. Unit. Nat. Devel. Progr., Food Agr. Organ., Rome. 71 p.

Leopold, A. S. 1959. Wildlife of Mexico. The game birds and mammals. Univ. Calif. Press, Berkeley. 568 p.

Man and the Biosphere (MAB). 1972. Final report. Expert Panel on Project No. 1. Ecological effects of increasing human activities on tropical and subtropical forest ecosystems in the Program on Man and the Biosphere. UNESCO, Paris. 35 p.

Nietschmann, B. 1971. The substance of subsistence, p. 167 to 181. *In* B. Lentrek, R. L. Carmin, and T. L. Martinson (eds.) Geographic research in Latin America—Benchmark 1970. Ball State University, Muncie, Indiana.

Nietschmann. B. 1972. Hunting and fishing focus among the Miskito Indians, Eastern Nicaragua. Human Ecol. 1: 41-67.

Nye, P. H., and D. J. Greenland. 1960. The soil under shifting cultivation. Tech. Comm. No. 51. Commonwealth Agr. Bureau Soil Sci. 156 p.

Ojasti, J. 1971. El Chiguire. Defensa de la Naturaleza 1(3): 4-14.

Ojasti, J., and G. Medina-P. 1972. The management of capybara in Venezuela. Trans. 37th. N. Amer. Wildl. Conf. p. 268-277.

Parker, I. S. C., and A. D. Graham. 1971. The ecological and economic basis for game ranching in Africa. Symp. Zool. Soc. London 24: 393-404.

Richards, P. W. 1952. The tropical rain forest, an ecological study. Cambridge Univ. Press, Cambridge, England. 450 p.

Sauer, C. O. 1958. Man in the ecology of tropical America. Proc. 9th. Pac. Sci. Congr. (1957) 20: 105-110.

Schulz, J. P. 1960. Ecological studies of the rain forest in northern Surinam. Noord Hollandsche Uitg. Mij., Amsterdam. 267 p.

Scudder, T. Ecological bottlenecks and the development of the Kariba Lake Basin, p. 206-235. *In* M. T. Farvar and J. P. Milton (eds.) The careless technology: Ecology and international development. Natural History Press, Garden City, New York.

Silver, W. S., Y. M. Centifanto, and D. J. P. Nichols. 1963. Nitrogen fixation by the leaf-nodule endophyte of *Tsychotria bacteriophyta*. Nature 199: 346-397.

Sternberg, H. 1968. Man and environmental change in South America, p. 413-445. *In* E. J. Fittkau, J. Illies, H. Klinge, G. H. Schwabe, and H. Sioli (eds.) Biogeography and ecology in South America. Vol. 1. Junk, The Hague.

Szestary, K. 1972. Hydrology and man-made lakes. Proc. Int. Symp. Man-made Lakes. University Tennessee, Knoxville.

Terborgh, J. 1973. Faunal equilibria and the design of wildlife preserves. Proc. II Int. Symp. Tropical Ecol. Caracas. Springer-Verlag, New York, in press.

United Nations. 1970. Integrated river basin development. United Nations, New York.

Went, F. W., and N. Stark. 1968. Mycorrhizae. BioScience 18: 1035-1039.

Section 6

IMPACTS OF REGIONAL CHANGES ON CLIMATE AND AQUATIC SYSTEMS

Charles Cooper, Team Leader

Pedro J. Depetris
James Ehleringer
Richard Fisher
Stuart Hurlbert
Stephen Schneider
Jay Zieman

I. Introduction

Man lives on a finite earth. Neither the full range of opportunities and alternatives nor the ultimate limits to his use of that earth are yet known. It is increasingly clear, however, that land, water, and atmosphere are intimately linked through complex biological and geophysical processes without respect to national political boundaries or even to continental limits. "Action at a distance" is indeed an accurate metaphor to describe the interregional and international impacts of one major ecosystem upon another.

Better understanding of regional ecosystem interactions can be of substantial help in private and governmental planning to maximize the contribution of natural resources to human welfare. For example, multidisciplinary studies of Lake Maracaibo, Venezuela, and its environs (Rodriguez, 1973) have shown that this highly industrialized tropical water body can be managed to maintain its biological productivity despite the perturbations of the oil and petrochemical industries. The oil industry in the Maracaibo basin yields a present economic return some 180 times that of the lake's fishery, yet the latter supports a significant number of people from lower income groups. The problems stem not just from oil pollution but also from complex changes in inflow and outflow patterns of lake water due to dam construction on its tributary streams and to navigation control structures at its meeting with the sea.

Rodriguez and his associates have shown how ecological analysis can suggest relatively minor changes in the operation of the dominant industry and its associated engineering works that will improve the fishery at relatively low cost. These changes involve not only the obvious actions to reduce the size and frequency of oil spills (their total elimination is virtually impossible in a lake producing more than three million barrels of oil daily), but also revision in placement and operation of water control dikes. Minor alterations in the plans for these dikes, yet to be built and intended chiefly to aid navigation, can result in current and salinity patterns benefiting rather than harming the fishery. Multidisciplinary ecological research can aid in reconciling industrial development and biological productivity.

Applied ecology in a regional setting involves four principal steps: (a) to identify the ecological perturbation and its cause, (b) to determine how to predict the long- and short-term consequences of the perturbation, (c) to apply the prediction to the formulation of regional plans, and (d) to define government action to enforce the regional plan and regulations.

Our concern here is with the first two of these steps. We have endeavored to identify major geophysical processes on a regional or continental scale which have significant biological implications. We have also attempted to define important actions of man likely to have a substantial ecological effect, beneficial or detrimental, on other regions or nations.

We first consider some of the major biogeochemical relationships in tropical freshwater, estuarine, and coastal ecosystems. Particularly important is the dependence of coastal and estuarine ecology upon terrestrial processes. Man's industrial and agricultural activities often upset natural relationships. The results are usually categorized under the broad headings of pollution and overexploitation. We give particular attention to problems of pollution in the near-shore marine environment. Questions of overexploitation of coastal fisheries, important though they are, have been treated extensively elsewhere (e.g., Rothschild, 1972) and will not be further considered here.

In reality, many widespread human diseases are ecological problems where regional land-water interactions are paramount. A case in point is schistosomiasis, which despite its importance in many tropical regions is uncommon in the Amazon Basin because of the local ecological context. Changes associated with accelerated development could, however, drastically alter the present favorable situation.

The worldwide circulation of the atmosphere rapidly carries material to distant regions. Little is known about prospects for alteration of tropical vegetation and soil through man-induced changes in atmospheric chemistry. Deliberate weather modification through cloud seeding may enable man to influence tropical precipitation amounts and hurricane intensities. This could have substantial ecological consequences. Of far greater long-term importance, however, is the possibility of global climatic change.

The region between 30° North and 30° South receives more than half of the incoming solar radiation driving the earth's climate. The possibility needs to be evaluated that man's activities, e.g., changing the turbidity of the atmosphere, altering regional evaporation patterns, modifying surface reflectivity, may generate large-scale and possibly irreversible changes in global climate.

Finally, advances in remote sensing and in computer simulation offer prospects for a major advance in our capability to predict, from regional climatic and soils data, the ecological capabilities of relatively remote regions, their resistance to stress, and their suitability for intensive human exploitation.

Perhaps most important of all regional ecological interactions, however, are those involving the flows of money, raw materials, finished products, people, and information between the tropics and the temperate zones. These problems have not been treated here, but the need for their study is acknowledged.

II. Recommendations for Research

1. Land and water are intimately linked through complex biological, geological, and geophysical processes. Interdisciplinary research is urgently needed to determine more fully the nature of these processes and the magnitude of their effects: (a) Expanded studies of the biogeochemistry of major tropical drainage basins are required to establish the initial physical, chemical, and biotic characteristics of their streams and associated estuaries. This information is needed as a basis for assessing the direction and magnitude of change under man's influence. Experimentation, including model building, should be undertaken to improve knowledge of the relative importance of environmental factors controlling the ecology and geochemistry of tropical rivers. Comparative studies of this sort should be started in relatively small watersheds, and then extended into larger tropical drainage basins. Hydrological, geochemical, and sedimentological studies should be carefully integrated with ecological research in the planning of such investigations. (b) Tropical estuaries have extremely low concentrations of nutrients and extremely rapid and conservative recycling mechanisms. The effects of added nutrients on these systems should be studied in more detail as the effects are apt to be different than in temperate estuaries. (c) The relative roles of various energy sources such as mangroves, algae, and river-borne detritus in tropical estuaries should be established, with particular emphasis on the effect of perturbation of these sources on food webs and fish yields. Similarly, there is a need to determine the detailed effects of alterations in fluvial regimes on the productivity of rivers, lakes, marshes, deltas, and estuaries. Such alterations may result

either from dam construction or other forms of river regulation, or from natural or man-induced climatic fluctuation. (d) Marshes, other wetlands, and running-water ecosystems in tropical regions are important to the ecology and food resources of distant lands as well as to the local area. Field studies should be undertaken to determine the quantitative importance of such systems for production and support of pelagic fishes and of waterfowl and other birds that migrate between continents and as habitat for resident birds and mammals.

2. Many currently recognized ecological problems are the consequence of large-scale human perturbations whose effects are felt at a distance. Among these are the transport of agricultural and industrial chemicals to distant estuaries, and engineering structures altering natural water flow patterns. (a) Additional research should be directed toward more fully establishing the acute and chronic limits of aquatic and estuarine organisms toward pollutants (e.g. pesticides, heavy metals), temperature, and wide fluctuations in ambient conditions, such as the effects of recurring droughts. Since it appears that many tropical aquatic and estuarine organisms normally exist closer to their tolerance limits than do most temperate counterparts, special attention should be given to rates of acclimatization of tropical organisms to changed conditions. (b) Hypotheses should be tested concerning the differential rates of food consumption, pesticide storage, excretion, and enzymatic degradation and the food chain concentration of pesticides in temperate and tropical organisms. The results will permit more valid extrapolation of existing temperate zone knowledge of pesticide-ecosystem relations to the tropics. (c) Expanded investigations should be made of tissue concentrations of organochlorine insecticides, PCBs, and selected heavy metals in carnivorous birds (pelicans, cormorants, hawks, owls, vultures) and of commercially important tropical fish species (both freshwater and marine). In regions where body loads of these pollutants are high, investigation of the possibility of declining reproductive success of these animal populations should be immediately initiated. (d) The influence of thermal and mixing regimes characteristic of tropical lakes and reservoirs on the circulation and biological effects of pollutants should be studied so that information that has been obtained in the temperate zone may help solve tropical problems. A similar comparison should be made of the nature of eutrophication in tropical versus temperate waters, fresh and marine. (e) Governments should require pesticide importers, distributors, manufacturers, growers, and international agencies to regularly submit exact information on importation, sales, uses, and use rates on a compound-by-compound basis. These data should be made public and published in conveniently accessible documents. Similarly, it is highly important that data on water quality be made readily available to the international scientific community. Where political or institutional barriers to the dissemination of the relevant

data are evident, scientists in the nations concerned should endeavor to point out the advantages of free exchange. (f) Before major modifications of shoreline physical conditions or of river flow regimes are undertaken, careful study be given to their probable effect on longshore current patterns, littoral drift, and sediment accumulation or erosion.

3. An intergovernmental committee should be established to explore the desirability, feasibility, and structure of a monitoring program designed to determine the sources and distribution of (a) organochlorine insecticides, (b) PCBs, and (c) selected heavy metals in a limited number of species widely distributed along the Pacific and Atlantic coasts of tropical Americas. Possible organisms might be (a) the sand crab *Emerita* sp. whose residue levels can serve as an excellent indicator of local pollution problems (see Burnett, 1971); (b) the mullet *Mugil* sp., which utilize estuaries and lagoons as nursery and feeding areas (they also move up and down coasts and their residues therefore might indicate pollution problems significant on a regional scale); and (d) the anchovy *Anchoa* sp. whose residues might serve as an index of pollution levels further off the coast. Such a monitoring program should be coordinated with any worldwide program that may be instituted under United Nations or other auspices.

4. Each coastal nation undertake a program to classify its estuarine resources on the basis of sound management and ecological considerations. Such a classification should delineate areas that should be reserved for heavy industrial development, those that are suitable for multiple use and those to be maintained as scientific preserves.

5. Many widespread human diseases can be dealt with most effectively as regional ecological systems in which land-water interactions are paramount and in which social, economic, demographic, political, agricultural, industrial, chemical, geological, and climatic factors all have strong influences. These disease systems should lend themselves well to a modeling approach. (a) Analysis of these disease systems should focus on large-scale regional aspects and particularly on how changes in regional demographic and land-use patterns affect aquatic ecosystems to the benefit or detriment of the aquatic vectors and hosts of disease organisms. (b) A critical review should be undertaken of the possibility that man will inadvertantly raise water pH, increase aquatic vegetation, and reduce populations of fish and other natural enemies of snails, accelerating the spread of schistosomiasis in the Amazon basin. Studies should be undertaken on water chemistry and dietary requirements of host snails, on their dispersal and migratory behavior, and on the role of natural enemies in controlling their populations. Information on these topics is prerequisite to realistic modeling efforts. (c) In many tropical regions, coordinated surveys should be taken of the present status of water-related public health problems, of aquatic populations (especially vertebrates), of river and lake hydrology, and of the physical-chemical characteristics of

the water in drainage basins presently being or soon to be subject to significant alteration by man. Such surveys can anticipate the kinds and magnitudes of problems that may develop and provide standards to measure and keep track of man's influence, favorable and unfavorable.

6. Increasing industrialization in tropical countries may well bring increased air pollution, particularly if high-sulfur fuels should predominate. Likewise, deliberate modification of weather for human benefit may eventually become feasible in the tropics. Both of these possibilities present important research problems: (a) An assessment of potential ecological consequences based on explicit consideration of local climate, plants and animals, land use, and patterns of human activity in the area of application should be made before initiating cloud-seeding programs to increase or decrease precipitation or to reduce hurricane intensity. (b) Additional research is needed to define the effects on native and introduced plant species of air pollutants generated by accelerated industrialization in tropical regions. (c) Additional research should be undertaken on the effects of large additions of sulfate on the chemistry of tropical soils and on the chemistry of heavy metals, both natural and as contaminants, in tropical soils.

7. Tropical ecosystems play a significant role in global climate, but the quantitative aspects of atmosphere-ecosystem interactions are largely unknown. Smoke and other particulate matter from tropical vegetation affect the turbidity of the atmosphere. Tropical ecosystems, particularly forests, help to regulate atmospheric carbon dioxide. Changes in vegetation cover may alter the global heat balance. Expanded collaborative research should be undertaken by ecologists and atmospheric scientists to obtain better quantitative data on ecological processes that influence climate: (a) Field research and simulation modeling are needed to determine more accurately the annual and seasonal evaporation and transpiration rates from extensive areas of various types of vegetation, natural and man-modified, in tropical lowlands. Only on the basis of such information can the magnitude and direction of any possible impact of major land-use changes on the global energy balance, and the resulting climate, be estimated with reliability. (b) Improved estimates of production rates, size distribution, and mean residence times of particles entering the atmosphere from burning of agricultural and forest wastes, including site preparation in shifting cultivation, should be made. Remote sensing at both thermal infrared and visible wavelengths is likely to prove useful in this respect. (c) Regular monitoring of the total biomass, and hence carbon storage of tropical forests, grasslands, and agricultural crops is needed. Multispectral remote-sensing methods are in prospect by which earth-resource satellite observations can be used to map the spatial distribution of biomass, chlorophyll, and leaf water.

8. The geographic scale and the large number of data points required for studies of ecosystem-atmosphere interactions virtually compel extensive use of numerical simulation and related mathematical methods. Ecologists should become more aware of current developments and applications in synthetic climatology and climatic modeling. Communication should be improved between ecological and atmospheric modelers. Many atmospheric scientists have little understanding of the problems inherent in situations lacking close relationships to physics, whereas biological scientists often feel that the parameters necessarily used by atmospheric scientists result in excessive oversimplification. Specifically, (a) Atmospheric scientists should be encouraged to continue efforts toward development of a spectrum of climatic models, including those that incorporate biological variables. Such models will permit more quantitative estimation of the possible impact on local and global climate of perturbations of the energy balance by modification of tropical ecosystems. (b) Atmospheric modelers and tropical ecologists should cooperate to devise a series of simulation experiments that can be performed with existing global climatic models. (c) Improved models of local and regional energy balance in the tropics should be designed to yield time-dependent estimates of potential and actual transpiration, soil moisture, transpiration deficit, and other ecologically significant variables. It may be desirable to base the models on synthetic as well as measured climatic data. Research should be undertaken to relate the resulting climatic estimates to observable characteristics of tropical ecosystems, using multivariate and other mathematical methods. This will result in improved procedures for expressing the environmental factors governing ecosystem structure and function, and for relating these factors to the observed properties of natural and man-modified ecosystems.

9. Modeling and prediction of ecosystem-atmosphere interactions will impose requirements for both climatic and ecological data on a greatly expanded scale. Therefore: (a) Cooperation should be improved between ecologists and atmospheric scientists concerned with global climatic measurements. Specifically, operations planners for the forthcoming Global Atmospheric Research Program (GARP) should be encouraged to incorporate ecologically important data into their program without jeopardizing the primary GARP objectives. For example, climatic measurements taken at a terrestrial location of interest to ecologists could be incorporated into the data net for the First GARP Global Experiment (FGGE), planned for about 1978. Interdisciplinary cooperation of this type implies establishment of an ecological advisory committee to GARP. (b) Designers of meteorological and climatological satellites should be encouraged to accelerate the development of methods for improved acquisition and retrieval of ecologically important climatic data by remote-sensing methods.

III. Ecological Impact of Technology on Aquatic Ecosystems

A. Biogeochemical Relationships

Rivers are the major pathways for transferring dissolved and suspended materials from the continents to the oceans. The introduction of these phases into the marine environment has sizeable impacts on the chemical composition of estuarine and coastal waters, and upon the concentrations of particulate matter in the sea and the mineral composition of sea-floor sediments. Clearly, the magnitude of the effect on the near-shore environment is a function of the fresh-water volume introduced by any particular drainage system. River influences have been observed hundreds of kilometers from the coast (Griffin et al., 1968; Goldberg and Griffin, 1970). Since tropical rivers are the largest source of fresh water entering the ocean (the Amazon River alone supplies 20 percent of all the river water discharged into the oceans) their impact on the marine environment is by far the most important.

The ecological characteristics of rivers are a function of the diverse interdependent environmental factors of their drainage basins, such as climate, biota, geology, relief, and soil types. Some of these factors are under variable stress induced by man's activities. Little is known about the role played by each variable in controlling biological and geochemical characteristics of tropical rivers.

Processes in rivers have their corresponding response in estuarine and coastal zones. In tropical regions, for instance, rivers transport large amounts of organic matter, such as floating vegetation and detritus. Most of this material moves during seasonal floods, when the plants are flushed from their natural environment, viz., ponds and marshes, into streams and rivers. This organic debris is largely responsible for the high productivity of estuaries.

Odum (1970) identified five important characteristics of the estuarine environment: (a) the nutrient-trap effect, (b) the unique structure of estuarine webs, (c) the harsh nature of the physical conditions and the resultant vulnerability of the estuarine organisms, (d) sedimentary control of estuarine waters, and (e) the key role of fresh-water inflow.

The nutrient-trap effect is due to a series of factors. The minerals carried into the estuary are usually deposited as the river widens and velocity decreases. In tropical estuaries the typical sedimentary discharges are clay-sized particles. The circulation of the estuary carries many of these particles out on the ebbtide and returns them on the flood. These clay particles have a high capacity to adsorb chemicals, including nutrients. Unfortunately this great sorptive capacity allows the estuary to act as a pollution trap, since pesticides and heavy metals are also adsorbed. At certain salinity levels, the electrical charges on the particles change and the nutrients—or pollutants—are released.

Tropical estuaries are usually biologically more diverse than their temperate counterparts, but they have very short food chains, which are primarily detritus-based as opposed to grazer-based. The organic detritus comes from river input from the bordering mangrove swamps and the input from tropical sea grasses, such as turtle grass *(Thalassia testudinum)*. The history of this plant illustrates the consequences of overexploitation and near extinction of a formerly abundant grazer and the conversion of a grazing food chain to a detrital system. Formerly, the vast turtle grass beds of the tropics were grazed by large herds of green sea turtles. Since the demise of these large herbivores, most of the production of this plant is not grazed and goes into the detritus food chain (Wood et.al., 1969).

The estuary is an extremely harsh environment for most organisms. It is an ecotone, a border community between the river (and land) and the sea, and is usually more productive than the river or sea. It has many of the properties of both neighboring systems as well as its own unique characteristics. It is a stressed environment, where physical parameters vary rapidly over short time spans and the organism must be adapted to handle these changes. It is, in the classification of Sanders (1969), a "physical-dominated" community, where the physical parameters and their variation are far more important in determining the communities present than classical biotic competition. These are important points since in the tropics biota live much closer to their lethal points than their temperate counterparts, with the result that small physical disturbances have a far greater proportional effect than in the temperate zone (see Moore, 1972).

The relative shallowness of many estuaries and the proximity of the sediments to the entire water column is partly responsible for high estuarine productivity. Bacteria in the sediments drive rapid regenerative mechanisms for the turnover of nutrients, such as compounds of nitrogen and phosphorus.

The fresh-water input sustains the salinity gradient defining the estuary. The middle zones of tropical estuaries are vital nurseries of many important animal species. Comparatively few organisms can survive in this section of the estuary, where conditions vary the most rapidly. Therefore, those organisms that can survive here, such as juvenile shrimp and mullet, face less competition for survival and less predation.

This brief review of the essentials of estuarine ecology, with emphasis on points of particular importance in the tropics, has shown that some mechanisms are quite susceptible to external alteration. Environmental changes often take place in concert, such as simultaneous changes in salinity and temperature regimes. Stressing an organism by modifying one parameter often lowers its tolerance for perturbations of all other parameters. This should be remembered in considering specific perturbations of the tropical estuarine environment.

B. Effects of Altered Water Regimes

1. Changes in flow

Seasonal flow patterns of rivers are mainly changed by the construction of dams and channels for irrigation and power generation. Significant changes in the watershed vegetative cover also have profound influences on the hydrologic behavior of streams and rivers.

The results of these man-induced changes are usually either a reduction or a sharp increase in the supply of silt and nutrients to downstream environments or a change in river dynamics. Both effects have profound impacts on biological activity in rivers, marshes, deltas, and other aquatic environments. Large areas in tropical river flood valleys, such as wetlands and ox-bow lakes, where highly productive ecosystems are located, are normally subject to an annual flood period. The amplitude and regularity of flooding is of the utmost importance in these "fluctuating water level ecosystems" (Odum, 1969), where communities are adjusted to the pulse of seasonal variations in water levels. Decreased water flow can also cause several pollutants, such as excess heat, particulate matter, industrial and domestic wastes, etc., to reach undesirable levels. Although we understand these specific effects, the rearrangement of hydrological patterns, such as the proposed creation of a series of man-made lakes that would interconnect the Orinoco, Amazon, and Rio de la Plata Basins in South America (Panero, 1967), will result in such large-scale changes that the ecological consequences cannot be predicted from current information.

The normal estuarine salinity distribution is a function of many factors, including tidal volume, circulation pattern, morphology, and fresh-water input. The variation of the fresh-water input to an estuary can produce profound changes on the biota. Reduction of fresh-water influx will cause the estuary to become more saline, advancing estuarine organisms into the formerly fresh-water zone. Effects on biota can be direct or indirect. When the fresh-water flow to the Everglades National Park was reduced, salinity increased in the estuary, and the area covered by mangroves increased 70 percent in less than 20 years. At the same time, the shrimp fishery in Florida Bay was damaged because juvenile shrimp were not able to detect the greatly diminished streams of low-salinity water emanating from the lower estuary. The young shrimp were therefore unable to find their optimal nursery grounds.

Changing the fresh-water inflow can also have profound effects by changing the stratification regime of an estuary. Reducing the fresh-water flow can allow more tidal mixing, whereas the increase of river-flow rates will tend to produce a more stratified, two-layered system. Obviously, a shift to a fresh-water regime will favor certain biota over others, reducing the ability of a community to withstand a "normal" drought (Heald, 1970). A less obvious change involving the true estuarine organism is that those inhabiting intermediate salinity water where rapid salinity change happens with each tide, will also shift distribution.

2. Changes in salinity

Changes in river flow can sometimes increase the salinity of river systems by concentration of natural dissolved loads by evaporation during the process of water recycling, or by raising the contribution of highly saline tributaries in the total flow of certain tropical drainage systems.

Salinity effects can also be subtle and cumulative. Lake Maracaibo was a fresh-water lake with a productive fishery. The opening of the navigation canal allowed the gradual accumulation of a layer of higher salinity at the bottom of the lake causing displacement of the normal fresh-water biota and the decline of the fresh-water fisheries. As noted in the introduction to this section, a number of engineering solutions to alleviate this situation and reduce the maintenance requirements of the navigation system are being considered.

Salinity changes of only a few parts per thousand usually have less effect on estuarine organisms than other variables (e.g., temperature) since variations in salinity are a part of the daily tidal fluctuation of the estuarine environments. Pronounced changes can be traumatic, however. Induced salinity changes are often accompanied by other undesirable environmental modifications, such as the effluents of a desalination plant, which are usually hypersaline (40 to 60 percent) and hot. Their increased density causes them to remain on the floor of the estuary with a large potential for ecological damage to the sessile organisms. In addition, the effluent may contain high concentrations of heavy metals. Sea-salt refineries introduce extremely hypersaline water (> 100 percent). Cintrón, Maddux, and Burkholder (1970) recorded stagnation, severe anoxia and accumulation of H_2S resulting from the effluent of a "salina" in Bahia Fosforescente, Puerto Rico. This caused extensive mortality of fish and estuarine invertebrates.

3. Thermal loading

Severe potential damage to the tropical coastal marine ecosystems is caused by increased thermal loading from the cooling effluent of electrical generating stations and certain industries. This is a particular problem in the tropics since tropical estuarine biota live much closer to their thermal maxima than their temperate counterparts (Moore, 1972). Biebl (1964) has shown that the thermal death point of marine algae in Puerto Rico is only 4 to 6°C above the average summer maximum whereas on the north coast of France the upper thermal limit is 10.5 to 13.5°C above the average summer maximum. Recent studies demonstrate that raising the temperature of tropical estuaries only 1 to 2°C can have a strong impact on the normal communities and the raising of the ambient only 3°C can destroy most of the biota of a subtropical estuary (Zieman and Wood, in press). Evidence is emerging that the optimal water temperature, even in tropic regions, is 30°C or below.

Increases in water temperature decrease the solubility of oxygen and therefore its concentration in water. At the same time the respiration of most biota is increased. This combination adds an additional stress to the system. In temperate regions, if temperatures are raised and the normal biota displaced, there is the possibility of replacement by more heat-tolerant forms from lower latitudes. In tropical estuaries there is no further source of higher heat-tolerant organisms tolerant to still higher temperatures. When the normal biota are destroyed by heat, they are usually replaced by heat-tolerant blue-green algae, which are sometimes directly toxic, or which may lead to ciguatera, an extremely severe poison concentrated in the higher trophic levels in marine tropical areas (de Sylva and Hine, 1970).

The vast columes of cooling waters of nuclear-power plants have reached the point where the circulation induced by the pumping of cooling water can overpower the normal estuarine circulation in areas where the tidal prism is small (Zieman and Wood, in press). This is not yet known to have happened in tropical estuaries, but may well occur as nuclear-power generation becomes more widespread. Because tropical marine forms are already near their thermal limits, there appears to be far less potential than in temperate regions for making constructive use of waste heat in mariculture or other schemes to increase biological productivity.

The discussion above has centered on marine and estuarine forms, as there is no information available on tropical fresh-water biota and their heat tolerance. It is to be expected however, that they will show the same problems of upper temperature limits as their marine and estuarine counterparts.

4. Suspended material

The increase of sediment load of rivers and estuaries is one of the major impacts of development. It has a variety of causes, and the effects of this increased sediment load on the biota of aquatic environments are many and varied. As development of the tropical river and coastal areas proceed, sedimentation will increase.

Suspended sediment concentrations on some major tropical rivers seems to be mainly controlled by relief. Gibbs (1967) showed that 82 percent of the total suspended sediments removed from the Amazon Basin was supplied by the Andean portion of the drainage. Rivers such as the Negro, Tapajos, or Xingu, with basins developed wholly within Amazonia lowlands, have very low suspended-solids concentrations and, hence, very low erosion rates—10.0, 3.8, 2.8×10^6 g km^{-2} yr^{-1} respectively, Nevertheless, given the tropical climatic conditions (including abundant and intense rains), certain man-induced changes in tropical ecosystems, such as deforestation, road construction, dredging and filling operations, and open pit mining, can significantly increase the suspended-solids concentrations of rivers even in the tropical lowlands where the controlling effect of relief is minimized.

Increased solid-phase concentration in river waters has marked biological and physical effects. Particulate matter or colored dissolved compounds such as humic acids, normally accompanying siltation, can diminish light penetration and hence the depth of the photic zone. This will bring nearer the surface the so-called compensation point, the point where production exceeds respiration. The decreased light levels may eliminate the highly productive macrophytes, such as the estuarine turtle grass *(Thalassia testudinum)* and also many species of macroalgae that require light intensities that do not normally occur below 10 m in tropical coastal zones. The net effect is a reduction in primary productivity and in food available for passage to higher trophic levels.

A marked decrease in the suspended-solids content of rivers resulting from impoundments can cause a reduction of the supply of silt and particulate organic matter responsible for the great productivity of deltas and estuaries.

Increased siltation modifies the geomorphological features of floodplains and rivers, adding sizeable amounts of sediments that must be removed by dredging to keep channels clear for river craft.

The biotic effects of the increased siltation in estuaries are often serious. Very high inputs of sediment kill biota by simply covering them. Corals can sweep away small amounts of silt, but are killed when they can no longer clear themselves. *Thalassia* and other sea grasses can recover from burial in moderate amounts (10 to 30 cm) of sediment by relying on stored starch reserves until they can photosynthesize again. Burial coupled with deteriorating water conditions causes the long-term destruction of the beds. Mangroves are also susceptible to sedimentary death. They possess aerial roots or pneumatophores which have air ducts leading to the deeper roots. When these ducts are clogged with sediment, the plants die within a few weeks.

The darker surfaces produced by turbidity also affect the heat distribution in water bodies with consequent effects on biological activity. When sediments are resuspended due to dredging or wind action they often release noxious gases, such as H_2S, and fine organic particles which have a high biological oxygen demand, noticeably reducing the dissolved oxygen content of the waters.

The effects of heavy-metal pollution will be discussed below. However, it should be kept in mind that metals are often complexed with silting particles. This has a variety of implications. In a sense it is a cleansing mechanism since the pollutants are removed from the water column. But even if the source of the pollution is stopped, the pollutant may be resuspended by the stirring of the sediments.

Littoral drift is responsible for the transportation of large volumes of sediments along the coastal zones. The interactions of this process with biological activity in coastal regions is insufficiently known. Nevertheless, sand transport along beaches can be substantially altered by man, either by a reduction of sand input caused by river damming or by physical modifications along the coasts (e.g., harbors, piers, coastal development, etc.). Ecologists working on coastal zones are often not fully aware of this phenomenon. Additional research is needed to foster sound advice on beach management and to alleviate existing problems.

5. Naturally occurring compounds

a. Dissolved solids. On a regional scale, atmospheric precipitation is the major mechanism controlling the chemical composition of low-salinity tropical waters, where the dilution by rainfall is high and the rate of supply of dissolved salts to the system is low. Tropical marine water may have natural nutrient concentrations lower than their temperate counterparts, but possess exceedingly rapid recycling mechanisms.

Gibbs (1970) has shown that the chemical composition and concentration of 16 tropical Amazon tributaries is close to that of the rainfall over the basin, and, with the exception of some rock-derived elements (such as silicon and potassium), it is also similar in composition to seawater. This clearly suggests the sea as the main source of salts transported into the precipitation cycle through wind systems.

All land-use practices that increase erosion rates in tropical river basins will accelerate the introduction of salts and nutrients into rivers, altering the river-water quality and eventually altering estuarine and coastal zones. Several agricultural practices linked with irrigation procedures, such as the fluctuating water level of wide use in rice crops, introduce varying amounts of salts through redox reactions, salts that would otherwise have remained in an insoluble state.

b. Nutrient enrichment—eutrophication. Construction of large reservoirs risks natural eutrophication of the lake and sometimes unpredictable and far-reaching consequences downstream, such as the impairment of the estuarine mechanisms controlling aquatic production. The growing use of fertilizers, which may be introduced into riverine systems by overland flow, accelerates river and lake eutrophication.

Eutrophicating compounds can be put into three general classes. The simplest are the fertilizers, which contribute only nutrients into the waters and simulate excess plant growth resulting in eutrophication. Certain industrial wastes such as those from sugar cane and copra processing plants in the tropics add nutrients, but also add organic solids, which immediately add to the Biological Oxygen Demand (BOD) load. These markedly decrease light penetration and oxygen concentration in water. The third group is sewage waste. This material has the properties mentioned above plus the added problem of adding potential pathogens into the water ranging from those that cause minor gastrointestinal discomfort to serious epidemic diseases.

Added nutrients, from whatever source, cause excess production leading to excess respiration at night. The resulting oxygen depletion may cause the destruction of some species. In addition, the communities favored by high nutrient levels are usually not composed of the same species as the natural community and may be noxious or even toxic. The vast quantities of excess organic material produced falls to the bottom. There, it combines with industrial and sewage wastes creating an anaerobic condition and blanketing the benthic communities. This is a serious form of pollution occurring in most harbors in the tropics. It presents one of the most immediate threats to man and his environment. Notable examples are Kingston Harbor in Jamaica (Wade, Antonio and Mahon, 1972), Hong Kong Harbor (Trott and Fune, 1973), and Guanabara Bay in Brazil (Castello, 1970).

6. Alien Pollutants

a. Heavy metals. Pollution from heavy metals comes from a variety of sources and is brought about by the release into the water of metals at a concentration from several times to several orders of magnitude above normal background levels. The sources are as varied as the metals themselves and include municipal sewage, agriculture, industrial wastes, and desalinization plants. Highly toxic metals include mercury, lead, cadmium, copper, nickel, chromium, and zinc. They present severe hazards because they may be cumulative in biota and are passed up food chains to higher trophic levels. They are also concentrated on particles and sediments up to 10^4 to 10^5 times their free-water concentrations and are not decomposed as some chemical pollutants are, but may be resuspended and liberated upon disturbance.

Mercury was the first heavy metal pollutant to which attention was drawn and is probably still the most important. In the tropics, mercury and other metals associated with biocides are apt to prove the most troublesome. Lead concentrations in estuarine waters of the Everglades due to agricultural runoff are the highest concentrations recorded in coastal waters (up to 200 μg/l).

Desalination-plant effluents release several tons of copper per year. This is entrapped in the stable dense effluent and absorbed into the benthic communities and the sediments causing severe problems.

With the increasing use of accurate and simple analytic techniques for the measurement of heavy-metal levels, such as atomic absorption techniques, the pervasiveness of a variety of metals is being discovered.

b. Oil. Minor oil spills are frequent in rivers with active traffic of river-going craft. Such spills can either occur during the transportation of oil and derivatives or during the cleaning process of bilges. Contrary to popular thought, the vast majority of oil released into aquatic systems comes from these normal industrial and transportation operations and not from the spectacular major spills.

Short-term acute effects (major spills) cause widespread destruction in the intertidal zone where the floating oil contacts the biota. Birds attract much attention because they are especially susceptible to damage from oil. Tropical mangroves also would appear to be very susceptible to damage by oil pollution. Rutzler and Sterrer (1970) observed that following an oil spill at Galeta Island, Canal Zone, seedlings of *Rhizophora* and *Avicennia* were killed when oil prohibited root respiration. A recent spill of aviation fuel, diesel oil, and bunker oil at the harbor entrance of Wake Island contaminated the coastline and killed over 2,500 kg of reef fishes and uncounted invertebrates (Gooding, 1971).

Compared with the severe acute spill, little is known of the effects of small quantities of long-term releases, or of the effects of the dissolving of certain soluble compounds from the oil into the water column. Oil at low concentrations has been shown to be deleterious to phytoplankton and marine invertebrate larvae. In addition to direct effects, oil is capable of being incorporated in the sediments and continues to be available for redispersal for indefinite time periods.

c. Pesticides. Tropical ecosystems are subject to the same effects of pollution by pesticides as are the temperate ecosystems from which most of our present information on these topics has been derived. Numerous reviews of this information have been made [see Study of Critical Environmental Problems (SCEP), 1970, and The Institute of Ecology, 1972 (MILE report), for global perspectives and summaries], and there is no need to repeat the generalities here. There are, however, certain specific features of the tropics, especially higher temperatures and greater predation, that directly influence the movements and effects of pollutants in tropical drainage systems. More research has been done on DDT than on any other major contaminant in the tropics. Therefore, despite indications that this compound is gradually being replaced by others of lesser persistence, attention in this discussion will be centered on DDT. Many of the principles developed are applicable, in either a general or specific manner, to circulation and effects of other pollutants in tropical drainage systems.

The ecology of DDT and related compounds has been extensively reviewed (Miller and Berg, 1968; Gillett, 1970; Study of Critical Environmental Problems, 1970; The Institute of Ecology, 1972), and significant additional studies have been published recently or are in progress. The severe effects of low concentrations of DDT on survival and reproduction of crustaceans, carnivorous fish, and carnivorous birds have been proven experimentally and are extensively confirmed by field studies. Data from field studies are less easily interpreted as our knowledge increases, however. There is now evidence that thin egg shells of carnivorous birds, for example, result not only from DDT, but also from other organochlorine insecticides, from PCBs, and from heavy metals such as lead (Witt and Gillett, 1970), and the levels of these various pollutants in the environment are often correlated.

Major uses of DDT in the tropics are in malaria control programs where it is applied to the interior of houses, in oncocerciasis control programs where it is appled directly to streams and rivers inhabited by blackflies *(Simulium)*, and in control of agricultural pests, especially in cotton. The claim that DDT applied to the interior walls of houses contributes negligibly to environmental contamination (Brown, 1970) seems reasonable but has been neither supported nor negated by empirical studies. The use of DDT for *Simulium* control is claimed to not cause fish mortality (Brown, 1970; Kershaw, 1966) but substantiating data are weak. Possible effects on fish reproduction have not been researched, and it is almost certain that deleterious effects are occurring on populations of fish-eating birds downriver from the treated localities.

Use of DDT in tropical cotton plantations bordering on lowland waterways and estuaries is frequently heavy, facilitating transfer of DDT from the agricultural to the aquatic systems. On the Pacific coast, "cotton plantations border approximately two thirds of the Guatemalan estuary system" and receive more than 100 pounds of insecticides per acre during the growing season (Keiser et al., in press). DDT is one of the principal pesticides employed, but toxaphene, heptachlor, parathion, and methyl parathion are also used, and most applications are pesticide mixtures (1964 data). Mean DDT residues of 36 to 45 ppm (whole-body basis) have been reported for two fish, *Mugil* sp. and *Poecilia sphenops*, which are eaten by the coastal human populations. It is suggested that DDT and other residues have decreased the estuarine nursery areas of the white shrimp *(Penaeus)*, on which another fishery is based. In some areas the local human population has been observed to eat fish apparently killed by dieldrin (S. Herman, personal communication). Studies of effects on tropical carnivorous birds are almost nonexistent, but Guatemalan cattle egrets are now producing extremely thin-shelled eggs (S. Herman, personal communication) and reproductive failures and thin egg shells have been reported for double crested cormorants on islands off off northwestern Mexico (Gress et al., in press). Shell thickness of these cormorants showed significant negative correlations with both DDT and PCB residues levels. No data are available on these problems in tropical raptors.

At the regional level perhaps the two major classes of ecological disruptions likely to be caused by DDT and similar substances are (a) destruction of fisheries, especially in the estuaries, but also in lakes and reservoirs, and (b) destruction of predator populations, both fish and birds, which may play crucial roles in regulating the structure and function of their ecosystems.

Crustaceans are extremely sensitive to DDT and many commercially exploited species live in estuaries or utilize them as nursery areas. In the tropics, fisheries also exist for fresh-water crustaceans such as *Macrobrachium*. Fish are less sensitive to DDT, but generally occur higher up the food chain and in this case may have higher DDT residues than their food. Well-documented cases of regional fisheries destroyed by pesticide pollution are rare even in the temperate zone, and nonexistent in the tropics. Declines in the speckled sea trout populations on the Texas coast and Dungeness crabs on the California coast are associated with the high DDT residues in eggs or larvae [Study of Critical Environmental Problems (SCEP), 1970]. Massive mortality in 1967 of the fry of the coho salmon from Lake Michigan was associated with high-residue levels in the parent fish (Johnson and Ball, 1972) and suggests that this new fishery may collapse as rapidly as it had developed.

In any region, lakes, reservoirs, and estuaries serve as DDT traps, greatly slowing the downstream movement of the pesticide. As discussed earlier, DDT entering one of these systems will be mostly adsorbed on particulate matter, much of which will either settle out as a result of slowed currents or will be incorporated into the food chains of reservoir organisms feeding on plankton or on particulate detritus. The creation of a reservoir on a river system contaminated by DDT can be expected to reduce the influx of DDT to downriver systems, including estuaries. Populations developing in the reservoir will, however, be subjected to higher DDT levels than were the populations in the former freeflowing river.

Tropical lakes and reservoirs, but probably not estuaries, show less frequent mixing of surface and bottom waters (turnover), higher rates of decomposition of organic matter, and longer periods of anaerobic conditions in the sediments and lower water column, than do similar systems in temperate zones. Major differences in the circulation patterns and degradation rates of DDT in temperate and tropical aquatic systems are expected. However, even for temperate systems hard data on these processes are scarce or nonexistent. Some attention has been given to degradation of DDT under anaerobic conditions (Fries, 1972), but the data are inconclusive for a comparison of anaerobic and aerobic breakdown rates.

The lower oxygen concentrations in tropical reservoirs also greatly prolong the time required for decomposition of the woody tissues of inundated vegetation, since the principal decomposers of lignin are aerobic fungi (Ruttner, 1963). Failure to clear forest in a developing reservoir basin may greatly hinder and delay the development of gill net fisheries on the reservoir, in which case concern over fish mortality or fish-residue levels become very academic matters from a fisheries point of view.

The rate at which individuals concentrate DDT residues in their bodies and therefore the rate at which DDT concentrations are magnified from one trophic level to another may be greater in the tropics than in temperate zones if respiration rates and food consumption rates are higher in the tropics. However, comparative data do not exist. If carnivorous species are more abundant in the tropics, if food chains are longer, and if second- and higher order carnivores occur more frequently, then not only is the number of species likely to suffer severely from DDT contamination much greater in the tropics, but also the "average" tropical carnivore will receive more residues and be more severely affected than will the "average" temperate zone carnivore.

Reduction of any single carnivore population in the tropics would usually have less effect than similar reduction of a corresponding temperate carnivore population because there may be a greater probability in the tropical system that other carnivores can increase to occupy the niche of the declining species. On the other hand, carnivory (or predation) appears to be a more important regulator in tropical than in temperate ecosystems, so a 10 percent reduction of carnivory (or of the pool of carnivorous species) might occasion a greater disruption of a tropical system than of a temperate one. As suggested in the previous paragraph, a given level of DDT pollution might be expected to cause the percent reduction of carnivory to be greater in a tropical than in a temperate system.

Principal disruptions to be expected from reduction of carnivore populations include increased amplitudes in the fluctuations of prey populations (and of any populations functionally linked to them), increased amplitudes in fluctuations of physical-chemical variables subject to biological influence (e.g., pH, turbidity, O_2 levels), minor to large shifts in mean values for population densities and for physical and chemical variables throughout the system, competitive exclusion or elimination in lower trophic levels by some species of other species with inferior competitive abilities but superior antipredator defenses (biochemical, morphological, behavioral). These types of disruption and species impoverishment will result not only from DDT but from any factor reducing carnivore populations. Nevertheless, with industrial and agricultural growth in tropical regions, the rate of environmental contamination by DDT, other organochlorine insecticides, and PCBs is likely to increase more rapidly than are the rates of other processes that deleteriously affect carnivores.

PCBs are chemically similar to many of the chlorinated hydrocarbon pesticides. They are found in many types of industrial effluents, and are highly toxic and pervasive. It seems that they are concentrated and transferred in similar ways to the pesticides, and like pesticides they are found in high concentrations in pelagic North Atlantic seabirds far from the sources of these compounds (Bourne and Bogan, 1972). No information is currently available about their distribution or impact in the tropics.

IV. Tropical Diseases as Ecological Systems

Many tropical diseases are ecological phenomena in which the interaction of terrestrial with aquatic ecosystems on a regional and local scale are of paramount importance. These include malaria, schistosomiasis (discussed earlier in Section 5), oncocercariasis, various forms of encephalitis, and other diseases where aquatic vectors or intermediate hosts are critical to the life cycle of the disease organism. As an example, consider the status of schistosomiasis (bilharziasis) in the Amazon basin and the likelihood of its further spread there. This example treats the largest river system in the world and the most widespread human disease caused by a metazoan parasite, clearly illustrating how man's use of the land can affect aquatic systems in a manner directly deleterious to himself.

At present, schistosomiasis is surprisingly uncommon in the Amazon Basin. The reasons appear to be the environmental requirements (especially pH and aquatic vegetation) of the *Biomphalaria* snails, which are obligate intermediate hosts for the schistosome *(Schistosoma mansoni)*, as well as certain cultural and demographic properties of the Amazonian human population. However, properties of the basin's aquatic environment and human population are changing rapidly as man develops the economic resources of the basin, and most of the anticipated changes are likely to favor an increased incidence and spread of schistosomiasis.

Where the *Biomphalaria* snails are absent, there is no risk of schistosomiasis, and snails appear to be absent in most Amazonian springs, creeks, and streams. These are rather acid (pH 5.5) and would tend to dissolve the calcium carbonate shells of any snails that did attempt to colonize them (Sioli, 1953). There are several ways in which man's activities may increase pH and remove this limitation. Lime applied to render the very acid soils of the Amazon region more suitable for agriculture could be leached into creeks and rivers, raising their pH. Although little lime presently is used in the Amazon Basin, its use is a prerequisite to agricultural development and will be increased as soon as limestone-crushing plants are established in the few good limestone deposits in the basin (Alvim and Araujo, 1952; Pinheiro, 1973).

Influx of nutrients in runoff from fertilized agricultural land or in human and animal wastes dumped into lakes, creeks, or rivers favors increased phytoplankton and macrophyte populations, which may raise pH values by photosynthetic uptake of CO_2. Brazilian agriculture so far makes very little use of commercial fertilizers, but its use doubled from 1962/64 to 1968 (Schuh, 1970). Impoundment of water in reservoirs also favors the development of phytoplankton and macrophyte populations, which might give a reservoir a higher pH than the water has flowing into it, once the inundated organic matter in the basin has been decomposed. Sioli (1967) reports pH values as high as 10 in naturally impounded flood-plain lakes of the Amazon.

Increased pH values in streams and rivers can also result from increased insolation of soils in altered terrestrial environments, e.g., as where natural forest is replaced by agriculture. Resultant higher soil temperatures foster accelerated decomposition of soil organic matter and reduction in numbers of soil microorganisms (possibly after a brief increase). Water percolating through such modified soil picks up smaller quantities of organic acids and smaller quantities of carbon dioxide (from microorganism respiration) than formerly, and the pH of the receiving streams will increase accordingly. In the Congo River Basin in Africa, Marlier (1972) has noted that streams flowing from denuded areas have higher pH (and higher salt concentrations) than do streams from areas of more intact vegetation, presumably as a result of the mechanism outlined above.

Snail distribution is also limited by availability of appropriate food, such as algae and aquatic vascular plants. These in turn, are limited by the extreme nutrient poverty of most Amazon waters and by forest canopies which overarch the streams to such an extent that even at midday less then one percent of the solar radiation reaches the water surface (Fittkau, 1964; Sioli, 1956). The significance of shade is seen at Fordlandia, an inoperative plantation established in 1928 when the natural forest was cleared right to the stream banks. Following this clearance, increased aquatic vegetation has confined to a good food supply for the *Biomphalaria* populations that subsequently established themselves in the more-or-less neutral water streams of the area. For at least the past two decades Fordlandia has served as the principal focus of schistosomiasis in the Amazon Basin (Sioli, 1956). Ordinances forbidding clearing of forests on banks of watercourses, such as those established by the Dutch in Indonesia as a siltation-preventive measure (Pelzer, 1961) might have a new and additional value in the Amazon system.

If agricultural and other activities of man tend to raise pH, increase dissolved nutrients, and increase the exposure to sunlight of presently snail-free streams, it is also true that the best soils for agriculture on the Amazonian "terra firme" are undoubtedly to be found in some parts of the Carboniferous bands along the Lower Amazon Valley," where the mostly neutral water creeks are already populated with *Biomphalaria*. This reflects the rough correlation of soil pH and the water pH and the fact that agriculture, like snails, fares poorly in a highly acid environment. It is thus conceivable that colonization and local population growth rates in the Amazon Basin will be greatest in just those areas where the possibility of schistosomiasis is greatest.

Other possibilities that merit attention are the reduction of fish, turtle, and other molluscivore populations by man, either inadvertantly by water pollution or as a result of his exploitation of these natural control agents for food. The introduction of the old world *Schistosoma haematobium* and *S. japonicum* by infected immigrants and the acceptability to these schistosomes of either native Amazonian or introduced snail hosts should be recognized as a very unlikely but also very dangerous possibility. After all, *S. mansoni* itself apparently is not native to the Americas but was introduced as a result of the African slave trade (Cheng, 1964).

Biomphalaria snails occur among the floating shore vegetation in flood-plain lakes and along the banks of several of the larger rivers (Solimões, Amazon, Madeira, etc.) in the basin. Although most of the basin's human population also lives along the banks of these same rivers, schistosomiasis has not become established in these areas. Sioli (1953) suggested as a partial explanation the facts that (a) the "Amazonian population does not have the custom of defecating in the water," (b) they avoid bathing in vegetated areas, and (c) a very high-dilution factor minimizes the probability of a schistosome miracidium encountering a snail. Central sewage systems emptying into watercourses and higher population densities might have unfavorable effects, if these explanations are valid. More information on human behavior and on snail population ecology along these larger rivers is clearly necessary before an understanding of the present situation and reasonable consideration of future prospects are possible.

V. Impact of Technology on Climate

A. Interregional Atmospheric Processes

Interactions between the atmosphere and the sea largely determine the chemical composition of precipitation. As already discussed, this strongly influences the ionic composition of tropical rivers. Pollution of the atmosphere by man may also become more important as industrialization progresses.

Smoke and haze from agricultural and forest burning already result in intense air pollution [see table 8.1 in Study of Man's Impact on Climate (SMIC), 1971] in tropical areas. Although such pollution may influence local and even regional climate, it has little direct effect on vegetation or soils. As tropical areas develop and industrialize, particulate matter in the atmosphere may be reduced because of decreased forest burning. But, many compounds that are now only trace components of the tropical atmosphere will be introduced in large quantities by industrialization.

Observations in temperate regions indicate that such emissions as sulphur dioxide, carbon monoxide, fluoride, ozone, lead and other heavy metals, and oxides of nitrogen become important and troublesome constituents of the atmosphere in industrialized areas. There is a highly developed symptomology for air pollutant effects on temperate vegetation and reasonable understanding of the susceptibility of many temperate plant species to various pollutants. Unfortunately, the effects of atmospheric pollutants on the survival and growth of tropical vegetation is virtually unknown. In addition, there seems to be no strong correlations between plant characteristics and susceptibility to air pollution which would allow us to predict the susceptibility of tropical species from temperate experience.

Large areas of vegetation in the vicinity of smelters and other heavy industries have been destroyed by air pollution in the temperate zone. This denudation has led to massive soil erosion, river and lake siltation, and the production of an altered environment that probably will never return to normal. If we are to avoid similar situations in the tropics we must not only use advanced methods to control industrial emissions but we also need to determine the susceptibility of tropical plant species to damage from atmospheric chemicals.

Another and somewhat more predictable effect of air pollutants is that of sulfur dioxide on soils. Low-cost high-sulphur-content fuels common in the early stages of industrialization lead to the production of large amounts of sulfur dioxide that reach the soil during rain storms as sulphuric acids.

The soils of the neotropics are quite diverse. They include very ancient soils developed on the Brazilian and Guayanan shields, somewhat younger soils of the Andean range, and very young soils of recent volcanic origin. As a result there is a wide diversity of nutrient status and a considerable range in pH (Beck and Bramao, 1968), Because the pH, buffering capacity, nutrient status, and content of weatherable minerals in the very old soils are already low, they would be much more easily damaged by additions of sulphuric acid than most temperate soils. One might expect the young volcanic soils to respond somewhat like their temperate counterparts. However, due to the accelerated rate of weathering in the tropics these soils might also be altered by acidic conditions.

Whereas the primary effects of sulphuric acid rains is to lower soil pH, there also are many secondary effects on soil chemistry. The solubility of most elements in the soil is quite dependent upon pH. In general, the availability of plant nutrients decreases with decreasing pH, whereas the solubility of many metallic elements increases. These changes could lead to great changes in leaching losses of elements from the soil. A change in the soils input of nutrient elements or metallic compounds to river systems could influence the aquatic biotic component.

Changes in soil pH might result in changes in surface and ground water quality, but this has not been conclusively demonstrated.

Although most tropical plants are adapted to grow under low pH and nutrient conditions, the extreme acidity and low nutrient status created by large inputs of sulphuric acids may lead to severe reductions in vigor and even death.

We have long considered that sulfur dioxide was the prime agent in this phenomenon; however, many sources of SO_2 also produce considerable quantities of heavy metals and these are often quite toxic to plants. Metallic elements remain in the soils long after the source of pollution has been removed and thus may have long-lasting deleterious effects.

B. Deliberate Weather Modification

Man may soon be able to exert a degree of deliberate control over local weather in the tropics. Artificial seeding of tropical cumulus clouds may increase or decrease precipitation over target areas by 20 percent or more. The ability to moderate the peak wind velocities of hurricanes, although probably not to steer them, may be achieved within a few years [National Advisory Committee on Oceans and Atmosphere (NACOA), 1972]. The low direct cost of weather modification technology, coupled with its apparent attractiveness for alleviating economic loss due to weather, ensures that there will be strong pressures for early application should weather modification become fully practical. Before this new technology is widely adopted, however, potential social, environmental, and economic costs as well as benefits should be addressed.

Several such assessments are under way in North America, Israel, and elsewhere, but their results will not necessarily be applicable to the American tropics. The only way to determine the ecological consequences of weather modification is to relate the specific way in which the weather regime will be altered to the existing climate vegetation, soil, land use, and patterns of human activity.

Deliberate weather modification largely depends upon introduction of artificial nucleating agents into properly selected cloud systems. Silver iodide is the most common nucleating agent, although several organic compounds may be more widely used in the future. There is a need for adequate evaluation of the long-term consequences of injecting potentially toxic materials into the atmosphere, but these materials are the same in temperate regions as in the tropics. Enough research on this topic is currently being done elsewhere so that research programs directed specifically toward determining the ecological effects of cloud seeding agents in the tropics are not warranted. In any event, preliminary evaluations indicate that silver contamination is not likely to be a serious problem, at least during the current exploratory stage in the development of weather modification technology (Cooper and Jolly, 1970).

There is considerable dispute about possible downwind effects, but the consensus among meteorologists is that if weather modification technology should prove capable of increasing or decreasing precipitation over a target area by as much as 20 percent, this will not result in a one-for-one change anywhere else. The precipitation scavenging process in the natural atmosphere is sufficiently inefficient such that artificial extraction of a little more moisture from a cloud system does not greatly change the amount available for precipitation downwind. Artificial cloud seeding on the scale now contemplated is not expected to alter cloudiness extensively, or to otherwise affect the factors influencing surface temperature (except to the limited extent that latent heat of evaporation of added or diminished rainfall affects local temperature). This would not necessarily be the case if worldwide modification schemes were mounted at a greatly expanded level.

This decision to initiate a rainfall-modification program will generally be motivated by the expected effect of the changed precipitation regime on agriculture or hydroelectric power. Since there is no way that the modification program can be confined to sharply delimited boundaries of land ownership or even national jurisdiction, an adequate assessment of its costs and benefits should include consideration of the potential effect on nontarget plants, animals, and people. This assessment will not be made easier by the fact that precise control over the amount of precipitation change cannot be achieved. It may be possible to determine statistically that precipitation over a season or a year has been increased or decreased by a given amount, within some probability range. Such a prediction cannot be made for an individual storm, however.

Decision makers may choose to alter precipitation throughout the year or only at certain seasons. Efforts may be concentrated on low rainfall years, when water shortages may be critical but few seedable air masses are around, or emphasis may be put on years with many suitable storms, when the same investment may yield a greater return of water but the water is less urgently needed. Innumerable combinations are conceivable. The ecological consequences of altering normal rainfall variability or its seasonal pattern may be more significant than those of altering the amount (Cooper, 1973). (But note in Section 5 the poor level of knowledge of normal rainfall.)

Research now in progress may lead to some degree of hurricane control. Hurricanes are among the most destructive of natural phenomena but also have some beneficial effects, particularly on the marine environment. They significantly affect the temperature of the ocean surface, and there is evidence of upwelling from a depth of 200 feet or more near the storm track. This can bring significant quantities of nutrient-rich water to the surface, thereby stimulating marine growth. Increase in phytoplankton standing crop near the surface to about double prehurricane values has been measured after a severe hurricane in the Gulf of Mexico. Whereas biological productivity at lower depths was reduced on account of turbidity restricting light penetration, the integrated effect was a substantial increase in the biological productivity of the water column (Franceschini and El-Sayed, 1968).

C. Global Climatic Change

Of greater significance for mankind than deliberate cloud seeding to alter rainfall is the possibility of altering global climate. The climate of the earth is maintained by a balance between the incoming solar radiation absorbed by the earth-atmosphere system within the outgoing planetary infrared radiation emitted to space by the earth's surface, the atmospheric gases and particles, and the clouds. The possibility of climate modification by *direct* human interference with the motions of the earth-atmosphere system is less likely than the chance of man's activities modifying the climate *indirectly*, by altering the energy balance or cloud nucleation processes. However, by diversion of rivers or ocean currents or by changes in the character of the land surface on a large scale, or by injection of carbon dioxide or particulate matter into the atmosphere, man might trigger mechanisms that affect the energy balance of the atmosphere—which could ultimately have an impact on climate.

1. Factors affecting the climate

The earth's atmosphere is a closed system, where at any particular time and place its state is related to and influenced by the atmospheric conditions at every other place on earth. The nonuniform latitudinal distribution of solar radiation reaching the globe and the heterogenous nature of the earth's surface combine to assure that the earth-atmosphere system is heated heterogeneously. Thus, each part of the system affects each other part because the *geographical differences* in heating drive the atmospheric heat engine.

The amount of solar radiation absorbed in the earth-atmosphere system must be equal over a sufficiently long time to the outgoing infrared or planetary radiation for the earth as a whole. However, this is not necessarily true locally. For example, the input of solar radiation in the equatorial latitudes exceeds the outgoing infrared radiation flux. This condition is referred to as a "positive" radiation balance. In middle latitudes, the incoming and outgoing radiant energies are about equal, whereas at the polar latitudes the outgoing infrared flux exceeds the absorption of incoming solar radiation by a large margin. The latter results from both the relatively small values of average incoming solar energy at the poles and the relatively high values of albedo (reflectivity) of the polar ice caps. This unequal or differential solar heating of the globe occurs primarily in zones (latitudinal belts). When coupled with the rotation of the earth, the unequal heating drives the motion we recognize as winds and ocean currents. These motions, both horizontal and vertical, regulate the distribution of temperature, cloudiness (which also requires the presence of suitable particles, called cloud nuclei), and precipitation over the globe, i.e., the major climatic variables.

The radiation balance of the earth-atmosphere system depends upon the concentration of optically active gases and particles in the atmosphere, the amount, kind, and thickness of the clouds, and the optical character of the earth's surface (especially the albedo or reflectivity of the surface to incoming solar radiation). Any alteration to these constituents persisting for a sufficient length of time (roughly 10 or more years), will alter the radiative heating of the earth-atmosphere system, and affect the thermal forcing of the atmosphere controlling the atmospheric motions.

The motions of the system result in a transport of heat from areas of positive radiation balance (the equatorial regions) to areas of negative radiation balance (the polar regions). In this process, the jetstreams, trade winds, eastward winds of the midlatitudes, westward winds of the polar latitudes, and migratory large-scale weather systems or eddies are all generated in the atmosphere. These variables, of course, estentially define the state of the climate. When the north-south temperature gradient is increased, the circulation becomes more vigorous in order to ameliorate the large temperature gradient. Eventually, the zonal nature of the east-west winds breaks down, generating large-scale transient eddies (storm systems) that transport additional heat and moisture poleward, reducing some of the equator-to-pole temperature difference. This situation is more characteristic of the winter hemisphere.

Thus, the temperature, winds, and precipitation in temperate latitudes are very much related to the same variables in the tropical latitudes. As an example, Namias (1972) has compiled statistical records showing a possible teleconnection between the highly variable year-to-year rainfall over northeast Brazil and the degree of cyclonic activity in the Newfoundland-Greenland area during the northern hemisphere's winter and spring.

The atmosphere carries heat in two forms—sensible and latent. For example, when warm air is directly transported to a cold region, it is a transport of "sensible" heat. Water vapor, evaporated or transpired at the earth's surface, can also be transported by the atmosphere. When it undergoes cooling in the presence of suitable nuclei (particles), the water vapor may condense into drops, releasing the "latent heat" needed originally to change it from liquid water to water vapor. Thus, any process that affects the evapotranspiration rate at the earth's surface or the concentration or character of the suspended atmospheric condensation nuclei that govern cloudiness and the release of latent heat will alter the energy balance of the earth's atmosphere system.

The process of evaporation, transport of water vapor, condensation, precipitation, and re-evaporation, which is called the "hydrological cycle," is responsible for about one-third of the net heat transported poleward across the 30° north and south latitude circles. Sensible heat transport by the atmosphere accounts for another third of the total heat transported, and oceanic currents such as the Gulf Stream carry the remaining third of the total heat flowing poleward. The ocean surface temperatures, controlled by turbulent mixing processes in the upper hundred meters or so of the oceans, play an important role in the exchange of sensible and latent heat with the atmosphere. Unfortunately, our knowledge of these processes is till inadequate to permit satisfatory quantitative determination of the oceans's role in the global heat budget.

2. Man's activities in the tropics that might affect climate

Human activities that might affect climate can be classified as modifications to the earth's surface, to the troposphere (i.e., lower atmosphere), or to the stratosphere (i.e., upper atmosphere). A detailed discussion of the entire subject of inadvertent climate modification can be found in the SMIC Report [Study of Man's Impact on Climate (SMIC), 1971]. SMIC concentrated heavily on the possible effects of industrial activities in the temperate latitudes, whereas use practices in tropical regions were given less attention. The general discussion in SMIC concerning the potential inadvertent consequences to the global climate from industrial activities in nontropical latitudes generally can be extrapolated to apply to industrial activities that are anticipated and current in tropical latitudes. However, since the region between 30° north and south latitude comprises half the earth's surface and receives considerably more than half of the incoming solar radiation absorbed in the earth-atmosphere system, it is apparent that large-scale modifications to either the land surface or atmospheric composition in tropical latitudes is likely to be more significant for global climate change than comparable changes in temperate latitudes. However, the climate in polar regions is probably the most sensitive to slight changes in the energy balance. Yet, it has been argued with the aid of simplified semiempirical mathematical models that the extent of the polar ice cover could be very sensitive to the radiation balance of the equatorial latitudes (Schneider and Gal-Chen, 1973).

Proposed tropical land-use projects, such as extensive alteration of the Amazon forests, would affect the physical character of a large and important fraction of the earth's surface. As we have stressed, these tropical regions are essential factors in governing the global energy balance and therefore large-scale modification projects might not have an effect limited to the local or regional climate, but might influence atmospheric processes thousands of miles away as well.

Change in either the albedo or moistness of the land surface could affect the local surface temperature. If the albedo of the surface were to be increased, leaving the moisture of the surface unchanged, then less heat would be absorbed at the surface, and the temperature would be decreased locally (assuming no local change in cloudiness resulting from the diminished surface heating). On the other hand, if the surface albedo were to remain unaltered while at the same time the moistness of the surface were changed, then there would be no change in the amount of solar energy absorbed by the surface, but there would be some change in the proportions of absorbed energy used to evaporate water, to that absorbed energy directly added to the local environment as sensible heat. If the surface moistness is increased while the albedo is unchanged, then more liquid water is evaporated at the surface and less sensible heat energy is available to warm the local region, and then the climate is cooled near the surface. At the same time, the evaporated water vapor is carried upward into the atmosphere and is transported downstream until it eventually condenses back into droplets in the clouds, releasing the latent heat energy needed originally to evaporate the liquid water at the surface and convert it to water vapor. Thus, there is a net energy loss at the original evaporation site (the surface) and an equal net energy gain at the condensation site (the atmosphere). Of course, the overall global energy input is unaltered since the albedo was unchanged.

If very large artificial bodies of water were to be constructed, they would likely warm the global climate and moderate the local climate, since it is probable that an artificial body of water will both decrease the surface albedo and increase the surface moistness. Deforestation, as explained below, is likely to do just the opposite. To examine this further, let us look at the geographic distribution and composition of diabatic heating of the atmosphere (bearing in mind that this heating drives the atmospheric "heat engine" or climate machine, as explained

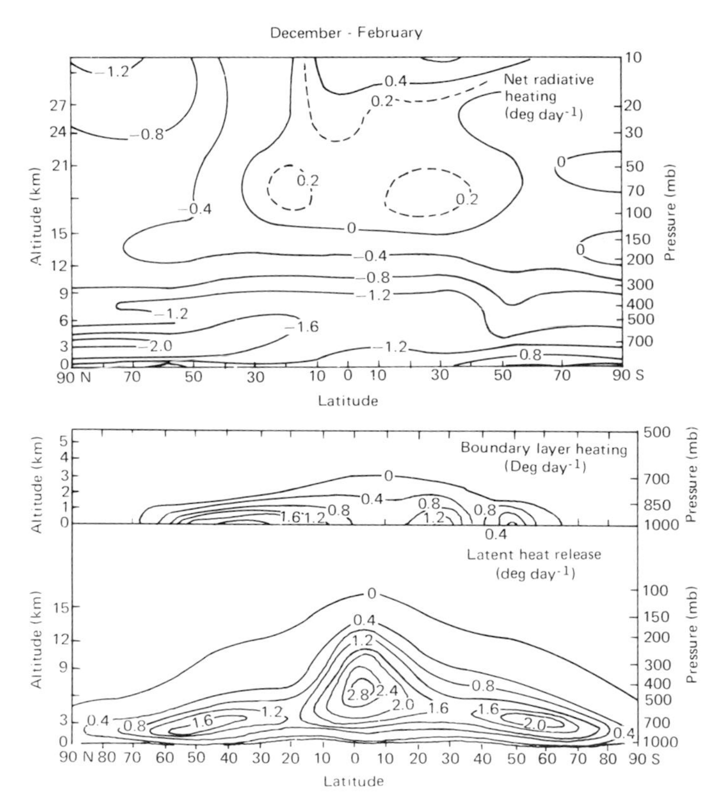

FIG. 6.1. Composition of diabetic heating for the atmosphere for December-February. From Newell et al. (1970.)

above). Figure 6.1 (from Newell et al., 1970) shows the heating of the atmosphere. Note the dominance of the latent release terms, especially in tropical latitudes.

Newell (1971) asks the question: "What will happen to the large-scale atmospheric general circulation if the tropical forests are removed over Brazil and Indonesia and perhaps over Central Africa? I think that the answer . . . is that we do not know, but we can speculate on the kind of influence it will have." Newell (1971) then speculates on the possible effects of deforestation: "Reduction of the water cycling between the forests and the air, which provides the latent heat high in the column, would clearly alter the pattern of the latent heat forcing function." If one examines the heat balance at say 500 mb and 5° south, one finds that the latent heat term is almost exactly offset by radiation and adiabatic cooling by rising motion (the magnitude of this is not shown in Figure 6.1, but must be about the same as the radiative cooling term). If a change occurs in the latent heat term one would expect changes in other terms also to maintain a balanced budget. It is difficult to even guess what they might be without a proper dynamical model. Boundary layer heating would obviously change due to the change in the surface albedo and conductivity as the forests are destroyed, although it should be borne in mind that the total longitude span over which the change of land surface occurs is small. Most of the contribution to the latent heat term comes from South America, Africa, and the Maritime Continent at this season, so that changing the tropical forests would interfere in a major way with this term. The radiational term would also change on account of less cloudiness.

Water vapor transport patterns, horizontal heat and momentum transport, and the convergence-divergence patterns of all these transports would also be expected to change. The amount of regular upper air information over the tropical continents is so small that we cannot yet establish the normal values of these transports properly.

Furthermore, as Newell points out, at low latitudes most of the latent heating is provided to the air in the regions of mean rising motion centered over South America, Africa, and the Indonesian region associated with summer season monsoonal circulations. Over the oceans between these land masses, mean sinking motions occur. Thus, the land areas (although comprising a smaller fraction of the total area of the earth's surface in tropical latitudes than the oceanic areas) are of more consequence in the heating of the global atmosphere than their relative surface area might suggest.

As an example of how strongly deforestation affects both evapotranspiration and runoff, we can consider the estimates from Sellers (1965, his Fig. 43) given here in Fig. 6.2. Although these data are for a middle latitude station (North Carolina) and may not be valid for tropical regions, deforestation can have a significant effect in decreasing evapotranspiration and increasing runoff as Newell has suspected. The extent to which similar differences would be observed in tropical regions would depend largely on the relative amounts of water available over the year to the root systems of native forests and to the crops that replace them. Field research and simulation modeling should be undertaken to determine the annual and seasonal evapotranspiration values from extensive areas of various types of vegetation

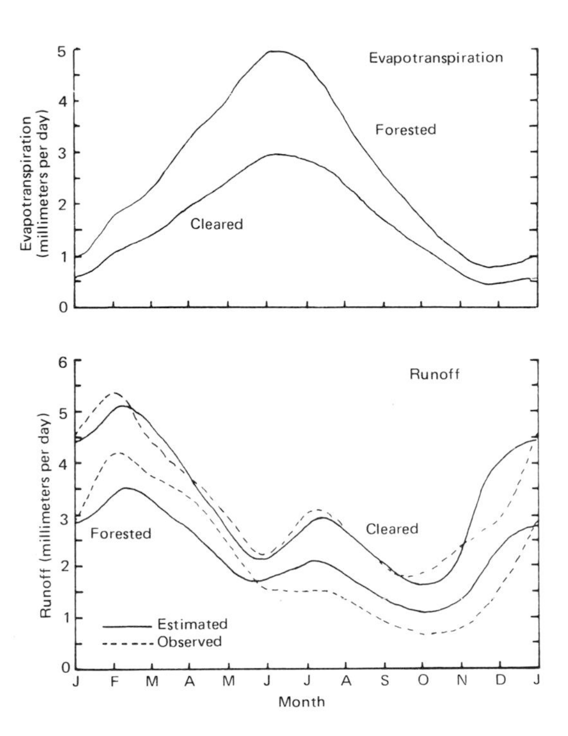

FIG. 6.2. Estimated evapotranspiration (top) and estimated and observed runoff (bottom) from cleared and forested watersheds near Coweeta, North Carolina. (From Sellers 1965.)

in tropical lowlands. Only on the basis of such information can the magnitude of any possible impact on the energy balance from tropical deforestation or other land-use schemes be quantitatively estimated with any degree of reliability. It is indeed not improbable that, at least in regions of heavy rainfall uniformly disturbed throughout the year, there would be little difference in transpiration between forests and cultivated crops.

Finally, it is essential to emphasize that while it is quite possible that land-use practices in tropical regions could significantly affect the local climate, the degree to which such climatic changes might be felt elsewhere is still entirely uncertain.

Although Newell (1971) has argued that tropical changes could affect the global energy budget, he has also recognized that predictions of any such possible effects must be approached with extreme caution. Newell states: "I should emphasize here that the linkages between the various parts of the general circulation are not well understood, even diagnostically. The extent to which latitude displacements of heat sources force changes in the tropics, and vice-versa, is particularly uncertain."

Thus, although we cannot "predict" precisely what climatic changes would result from large-scale changes in usage, we can say that a modification to the energy balance much larger than a few tenths of 1 percent would be expected to cause climatic changes both locally and in other parts of the world (Budyko, 1972). Thus, it is important to estimate the magnitude of the effects on the energy balance of agricultural development on the scale proposed. It is necessary to determine how such a perturbation to the energy balance might affect the climate. This exercise presupposses the existence of a quantitative theory of climate, expressible in the form of mathematical models of the earth atmosphere system. Such models must be complete enough to include all important climatic feedback mechanisms (such as changes in cloudiness or polar ice that might be associated with the change in energy balance which could "feed back" and either ameliorate or accentuate the initial effect on the climate of the change in the tropical energy balance that followed the deforestation). Although current climate modeling efforts are still insufficient to determine confidently the ultimate effect of tropical land-development schemes on the climate, several existing models include many of the important atmospheric feedback mechanisms, and should be applied to this question in order to produce a first estimate of the possible effect.

For a more complete description of climatic modeling and feedback mechanisms see Chapter 6 of the SMIC Report or the survey article by Schneider and Kellogg (1973).

Slash and burn agricultural practices, resulting in the injection of a considerable amount of dust and smoke into the atmosphere, is another example of a human activity in tropical ecosystems that could also have inadvertent climatic consequences. As stated earlier, particles in the atmosphere can affect the energy balance of the earth-atmosphere system by two processes: (a) by directly altering the radiation field by scattering and absorbing radiation, and (b) by providing condensation nuclei that can affect the cloud forming processes, with possible consequences on precipitation, latent heat release and albedo (reflectivity) of individual clouds. (This subject is treated in considerable depth in Chapter 8 of the SMIC Report.)

The atmospheric particle loading is a result of both natural and man-made processes. Windblown soil and rock debris, evaporated sea water that leaves sea salt particles in the air, organic exudation from plants that forms particules, and sulfate or nitrate particles (which are originally in the form of oxide gases of sulfur or nitrogen that are subsequently photochemically converted to particles) are all examples of naturally produced particle "pollution." Of course, deforestation or desertification resulting from man's activities will substantially increase the amount of windblown soil and rock debris, and it is fair to question whether such a particle source is really "natural."

Man-made particle sources result primarily from the photochemical conversion to sulfate particles of sulfur dioxide gas associated with the burning of fossil fuels. Also, there is a measurable direct injection of man-made particles into the atmosphere. However, the most important source of man-made particles could be those that result from forest fires and slash and burn agricultural practices. Unfortunately, estimates of this source range from 3×10^6 all the way up to 150×10^6 metric tons per year (see Table 8.1 of the SMIC Report), a spread in uncertainty which is highly unsatisfactory for determining the possible influence of vegetation fires on the global energy balance. SMIC shows that our present knowledge of the sources and sinks of atmospheric particles is such that we can only estimate at present that the man-made fraction of the total particulate loading is from 5 to 50 percent. Clearly, improvement of these estimates should be an important research priority.

Remote sensing at both thermal and infrared wavelengths may be useful in such improvement. Thermal imagery from earth resource satellites can quite easily display the number and area of clearance fires burning at any one time. Spectral analysis should make it possible to differentiate such fires from other heat sources. Imagery at optical wavelengths can be used to determine the relative area of burned vegetation and the extent of smoke plumes. These data can, in turn, be used to estimate the total input into the atmosphere as a function of time.

Remote sensing also seems likely to permit mapping of the spatial distribution of biomass, chlorophyll, and leaf water in tropical and other regions. Techniques have been developed for determining the biomass of large areas (ca. 1.9 ha) of grassland vegetation from multispectral scanner data. Correlations of 0.8 to 0.9 between estimated biomass values from multispectral data and from clipped plots have been obtained (Pearson and Miller, 1972). It appears likely that these techniques can be extended to forest and shrub vegetation. If so, the way would be open to more accurate estimation of the role of tropical vegetation in regulating the global carbon dioxide balance, as well as to a variety of other ecological analyses.

In summary, a plausible argument can be made to the effect that alteration of tropical ecosystems by land development or increase in slash and burn practices can alter the energy balance of the earth-atmosphere system enough to cause more than local changes in the climate. While it is difficult at present to be very precise about what kind of changes might occur and where, a continued program of development of mathematical models of climate change will undoubtedly help our understanding of any possible consequences to the climate arising from use practices. At the same time, the magnitude of environmental stress from such practices needs to be estimated with far more precision than has been done at present.

There is considerable evidence (see SMIC Report, Chapter 6) that if changes to the global climate are allowed to proceed, they could then become irreversible. It is essential that more definitive answers to the questions posed about inadvertent climate modification resulting from man's activities be found before the changes are likely to occur. We now have a unique opportunity to evaluate the possible impact of large-scale technological efforts before the projects are completed. In view of the negligibly small expenditures that this sort of research effort will cost in relation to the vast sums to be spent on the large-scale projects, a continous effort at understanding the consequences of modifications to terrestrial ecosystems seems truly essential.

3. Atmospheric oxygen: a nonproblem

It is a usual question to ask what might happen to the concentration of life-sustaining atmospheric oxygen should there be a reduction in photosynthetic organisms, either through herbicide destruction or by direct deforestation practices. However, recent measurements and calculations (see SCEP Report, p. 74-75) show that there would be virtually no change in the concentration of oxygen for tens of thousands of years, even if, for example, all the known recoverable fossil fuels were burned. Of course, the carbon dioxide released by such an action would be of considerable importance to the global climate, but atmospheric oxygen is essentially a nonproblem.

D. Synthetic Climatic Statistics As Ecological Predictors

New possibilities have opened in recent years for more productive research on the regulatory role of climate in ecosystem structure and function. Advances in remote sensing and in computer simulation models of global atmospheric circulation promise that greatly improved representations of ecologically important climatic variables can be determined for tropical regions. This in turn would lead to a major advance in our capacity to predict the ecological capabilities of relatively remote and unknown regions, their resistance to stress, and their suitability for various kinds of human exploitation.

If the promise of the new technological methods is fulfilled through research, it should be possible to present, for any desired terrestrial location, synthetic arrays of climatic variables that preserve the statistical structure, although not the instantaneous values, of the actual climatic conditions. This would mean that simulated regimes of temperature, solar radiation, atmospheric humidity, wind velocity, and eventually perhaps precipation, can be produced for places lacking adequate long-term weather records. These simulated regimes would have seasonal patterns, means, variances, and other statistical properties similar to those of the real world climate, even though they were not derived from ground level measurements of the weather at the location being considered.

Much effort has been devoted to estimating unmeasured characteristics of tropical physical environments from observed physiognomic and structural characteristics of the vegetation. Such indirect estimation has been necessary because of the paucity of adequate environmental measurements. Examples are the vegetation classification scheme developed by Webb (1968) in Australia, and the Holdridge Life Zone System, widely used in Latin America. The possibility now exists that this traditional approach to environmental analysis may be entirely reversed. Instead of attempting to infer the nature of the environment from its vegetation, it may soon be feasible to specify the physical environment almost completely. This specification can then be used to better explain and predict the characteristics of a region's native ecological communities.

Contemporary satellite systems permit statistical studies of cloud cover. Near-earth polar satellites provide a nearly global picture of cloud cover twice every day. There are plans for advanced instruments to be flown in the 1970s which will make it possible to determine a coarse vertical profile of total water vapor in the atmosphere with an accuracy of about 20 percent. Instruments capable of determining the surface temperatures to better than 1°C absolute with about 0.25°C precision are planned for forthcoming meteorological satellites. A combination of available and projected satellite techniques is expected to permit estimation of net radiation at the surface. There are, however, at this time no satellite developments that might provide satisfactory observing techniques for local rainfall [Committee on Space Research (COSPAR), 1972]. A range of possible statistical and other estimating methods may be usable for this purpose by the time required climatological satellites are fully operational, however (Follansbee, 1973).

Concurrent with development of satellite observational techniques, work has progressed on the design of large computer simulation models of the global circulation of the atmosphere, particularly at the National Center for Atmospheric Research (NCAR) and at the National Oceanographic and Atmospheric Administration (NOAA) Fluid Dynamics Laboratory of the United States Department of Commerce. These models are becoming increasingly effective in mimicking the annual patterns of significant climatic variables everywhere. They can be operated to simulate a year's climate of the whole earth, at a scale of resolution of 5° or less of latitude and longitude, in a small fraction of a day. Repeated simulation runs of the global circulation models can be used to build up statistical distributions of climatic variables of interest, in effect extending in time the short period of record obtainable from satellites. There is now close collaboration between researchers concerned with obtaining and improving satellite observations and those developing atmospheric circulation models.

Synthetic climatic data will permit computation of regional energy balance and evapotranspiration values. The Penman method for determination of potential transpiration, as modified by van Bavel (1966) and others, provides a starting point for this investigation. Given regional values of net radiation, temperature, humidity, daily wind movement, and moisture holding characteristics of the soil, relatively simple computer programs can be written to estimate potential transpiration, actual transpiration, transpiration deficit, and soil-moisture storage. These variables can be summed over convenient periods, say half-months, and displayed as vectors representing the annual march of climatic phenomena. Associated with each vector element can be a measure of its variance from year to year. This sort of multidimensional representation of the climatic variables that drive ecosystem processes should logically yield more reproducible and interpretable results than would simple correlations of vegetation structure and function with annual rainfall and temperature alone. There will, of course, be a need for a number of carefully chosen calibration and validation sites, to test the methods in the field and to provide data for "tuning" the models.

Multivariate methods now exist for relating vectors such as transpiration and soil-moisture sequences to observable characteristics of biological systems. The eventual goal should be a scheme for arranging ecosystems in relation to one another on a hypersurface built around a biological model of responses to amounts, time distributions, and variances of soil moisture, temperature, solar radiation, and other pertinent environmental variables.

Development of satellite sensors capable of making the relevant observations has largely been motivated by a desire to understand and predict large-scale processes and events in the atmosphere itself. The prospective capability to make surface measurements of the type needed by ecologists is an almost accidental consequence of atmospheric scientists' needs for the same kinds of data. There is a need for greatly improved cooperation between ecologists and those atmospheric scientists concerned with global climatic measurements. Specifically, designers of meteorological and climatological satellites should be encouraged to accelerate the development of methods for improved acquisition and retrieval of data likely to be of value to ecologists as well as to atmospheric scientists.

There is a danger here that must be clearly faced, however. For the foreseeable future, meteorological and earth resource satellites will be operated in any numbers only by the U.S. and the U.S.S.R. As the Swedish ecologist Bengt Lundholm (1972) points out, relative to remote sensing we are in a situation where the technical development is much more advanced than the political tools with which to handle this development. He states

> A certain sensitivity, to use an understatement, may be felt by nations overflown by satellites that make inventorial surveys for economic reasons. Information concerning the agricultural aspects of natural resources might be used against the overflown country, especially if this type of information is not available to everyone. According to the USA's plans for earth-surveying satellites, all information is to be open and free. In spite of this there is a real danger that this new technology will increase the gap between the developed and the less developed countries.

There have already been objections from certain nations to the open dissemination of information discovered by satellite that may indicate previously unknown mineral deposits in their national territory. Considerable attention will be required to ensure that proper institutional arrangements are made for effective use of remote-sensing information in the general interest.

Somewhat similar conditions exist in instances where only a few nations possess the powerful computing systems that are required for truly large-scale atmospheric, geophysical, and ecological models. Institutional mechanisms should be developed to ensure that scientists from the countries or regions concerned are closely involved in the planning and conduct of computer simulation experiments involving ecological relationships in their countries.

References

Alvim, P. de T., and W. A. Araujo. 1952. El suelo como factor ecologico en el desarrollo de la vegetacion en el centro-oeste del Brasil. Turrialba 2: 153-160.

Beek, K. J., and D. L. Bramao. 1968. Nature and geography of South American soils. p. 82-112. *In* Biogeography and ecology in South America. V. I. Junk, The Hague, Netherlands.

Biebl, R. 1964. Temperaturresistenz tropischer Meeresalgen. Bot. Marina 4: 241-254.

Bourne, W. R. P., and J. A. Bogan. 1972. Polychlorinated biphenyls in North Atlantic seabirds. Mar. Poll. Bull. 3: 171-175.

Brown, A. W. A. 1970. The present place of DDT in world operations for public health, p. 196-197. *In* J.W. Gillett (ed.) The biological impact of pesticides on the environment. Proc. of Symposium. Oregon State University Press, Corvallis, Oregon.

Budyko, M. I. 1972. The future climate. Trans. Am. Geophys. Union 53: 868-874.

Burnett, R. 1971. DDT residues: Distribution of concentrations in *Emerita analoga* (Stimpson) along coastal California. Science 174: 606-608.

Castello, O. O. 1970. Pollution in Guanabara Bay, Brazil. Mar. Poll. Bull. 1: 150.

Cheng, T. C. 1964. The biology of animal parasites. Saunders, Philadelphia, Pennsylvania. 727p.

Cintrón, G., W. S. Maddox, and P. R. Burkholder. 1970. Some consequences of brine pollution in the Bahia Fosforescento, Puerto Rico. Limnol. Oceanogr. 15: 246-249.

Cooper, C. F. 1973. Ecological opportunities and problems of weather and climate modification, p. 99-134. *In* W. R. D. Sewell (ed.) Modifying the weather: a social assessment. Western Geographical Series, V. 9. University of Victoria, Victoria, B.C.

Cooper, C. F., and W. C. Jolly. 1970. Ecological effects of silver iodide and other weather modification: A review. Water Resources Res. 6: 88-98.

Committee on Space Research (COSPAR). 1972. Preliminary observing system. Considerations for monitoring important climate parameters. COSPAR Working Group 6, Boulder, Colorado. 87p.

Fittkau, E. J. 1964. Remarks on limnology of central-Amazon rain-forest streams. Verh. Internat. Verein. Limnol. 15: 1092-1096.

Follansbee, W. A. 1973. Estimation of average daily rainfall from satellite cloud photographs. Nat. Ocean. Atmos. Admin. Tech. Memo. NESS 44.

Franceschini, G. A., and S. Z. El-Sayed. 1968. Effect of Hurricane Inez (1966) on the hydrography and productivity of the western Gulf of Mexico. Deutsche Hydrograph. Zeit. 21: 193-202.

Fries, G. F. 1972. Degradation of chlorinated hydrocarbons under anaerobic conditions, p. 256-270. *In* S. D. Faust (chmn.) Fate of organic pesticides in the aquatic environment. Amer. Chem. Soc., Washington, D. C.

Gibbs, R. 1967. The geochemistry of the amazon river system: Part I. The factors that control the salinity and the composition and concentration of the suspended solids. Geol. Soc. Amer. Bull. 78: 1203-1232.

Gibbs, R. V. 1970. Mechanisms controlling world water chemistry. Science 170: 1088-1090.

Gillett, J. W. (ed.). 1970. The biological impact of pesticides in the environment. Oregon State University Press, Corvallis. 210p.

Goldberg, E. D., and J. J. Griffin. 1970. The sediments of the northern Indian Ocean. Deep-Sea Res. 17: 513-537.

Gooding, R. M. 1971. Oil pollution on Wake Island from the tanker *R.C. Stoner*. Nat. Mar. Fish Serv. S.S.R. #636.

Gress, F., R. W., Risebrough, D. W. Anderson, L. F. Kiff, and J. R.

Jehl, Jr. 1973. Reproductive failures of double-crested cormorants in southern California and Baja California. Wilson Bull., in press.

Heald, E. J. 1970. The Everglades estuary: An example of seriously reduced inflow of freshwater. Trans. Amer. Fish. Soc. 99: 847-850.

Johnson, H. C., and R. C. Ball. 1972. Organic pesticide pollution in an aquatic environment, p. 1-10. *In* S. D. Faust (chmn.) Fate of organic pesticides in the aquatic environment. Amer. Chem. Soc., Washington, D. C.

Keiser, R. K., Jr., J. A. Amado A., and R. Murillo S. 1973. Pesticides levels in estuarine and marine fish and invertebrates from the Guatemalan Pacific coast. Univ. Miami Bull. Mar. Sci., in press.

Kershaw, W. E. 1966. The *Simulium* problem and fishery development in the proposed Niger Lake, p. 95-97. *In* R. H. Lowe-McConnell, (ed.) Manmade lakes. Academic Press, New York.

Lundholm, B. 1972. Remote sensing and international affairs. Ambio 1: 166-173.

Marlier, G. 1973. Limnology of the Congo and Amazon Rivers, p. 223-238. *In* B. J. Meggers, E. S. Ayensu, and W. D. Duckworth (eds.) Tropical forest ecosystems in Africa and South America: A comparative review. Smithsonian Institute Press, Washington, D.C.

Miller, M. W., and G. G. Berg (eds.) 1969. Chemical fallout: Current research on persistent pesticides. Thomas, Springfield, Illinois. 531p.

Moore, H. B. 1972. Aspects of stress in the tropical marine environment. Adv. Mar. Biol. 10: 217-269.

National Advisory Committee on Oceans and Atmosphere (NACOA), 1972. Report to the President and to the Congress. National Advisory Committee on Oceans and Atmosphere, Washington, D. C.

Namias, J. 1972. Influence of northern hemisphere general circulation on drought in northeast Brazil. Tellus 24: 336-343.

Newell, R. E. 1971. The Amazon forest and atmospheric general circulation, p. 457-459. *In* W. H. Matthews, W. W. Kellog, and G. D. Robinson (eds.) Man's impact on the climate. MIT Press, Cambridge, Massachusetts.

Newell, R. E., D. G. Vincent, T. G. Dopplick, D. Feruzza, and J. W. Kidson. 1970. Energy balance of the global atmosphere. Global circulation of the atmosphere. Roy. Meteorol. Soc., London.

Odum, E. P. 1969. The strategy of ecosystem development. Science 164: 262-270.

Odum, W. E. 1970. Insidious alteration of the estuarine environment. Tran. Amer. Fish. Soc. 99: 836-847.

Panero, R. B. 1967. A South American "Great Lakes" system. Preliminary Report HI-788/3-RR. Hudson Institute, New York. 29 p.

Pearson, R. L., and L. D. Miller. 1972. Remote mapping of standing crop biomass for estimation of the productivity of the shortgrass prairie, Pawnee National Grasslands, Colorado Proc. Eighth Int. Symp. on Remote Sensing of Environment, October 2-6, 1972, Ann Arbor, Michigan Environmental Research Institute of Michigan, Ann Arbor, Michigan.

Pelzer, K. J. 1961. Land utilization in the humid tropics: Agriculture. Proc. 9th Pacific Sci. Congr. (Bangkok, 1957) 20: 124-143.

Pinheiro Codurn, J. M. 1973. Agriculture in the Amazon. In Proc. 23rd Ann. Latin Amer. Conf. "Man in the Amazon". University of Florida, Gainesville, in press.

Rodriguez, G. 1973. El Sistema de Maracaibo. IVIC, Caracas, in press.

Rothschild, B. J. (ed.) 1972. World fisheries policy. University of Washington Press, Seattle. 272p.

Ruttner, F. 1963. Fundamentals of limnology. University of Toronto Press, Toronto. 295p.

Rutzler, K., and W. Sterrer. 1970. Oil Pollution. Damage observed in tropical communities along the Atlantic seaboard of Panama. BioScience 20: 222-224.

Sanders, H. L. 1969. Benthic marine diversity and the stability-time hypothesis, p. 71-81. *In* Diversity and stability in ecological systems. Brookhaven Symp. Biol. Vol. 22. Brookhaven National Laboratories; Upton, New York.

Schneider, S. H., and T. Gal-Chen. 1973. Numerical experiments in climate stability. J. Geophys. Res. 78: 6182-6194.

Schneider, S. H., and W. W. Kellogg. 1973. The chemical basis for climate change, p. 203-249. *In* S. I. Rasool (ed.) Chemistry of the lower atmosphere. Plenum Press, New York.

Schuh, G. E. 1970. The agricultural development of Brasil. Prager, New York. 456 p.

Sellers, W. D. 1965. Physical climatology. University of Chicago Press.272p.

Sioli, H. 1953. Schistosomiasis and limnology in the Amazon region. Amer. J. Trop. Med. Hyg. 2: 700-707.

Sioli, H. 1956. The only focus of bilharziasis *(Schistosoma mansoni)* in the Amazon region. Observations on the ecology of the vector *Tropicorbis (Obstructic) paparyensis* F. Baker and their practical importance. WHO/Bil. Ecol. 1, 23 May 1956.

Sioli, H. 1967. Studies in Amazonian waters. Atlas do simposio sobre a biota Amazonica 3: 9-50.

Study of Critical Environmental Problems (SCEP). 1970. Man's impact on the global environment. MIT Press, Cambridge, Mass. 319p.

Study of Man's Impact on Climate (SMIC), 1971. Inadvertent climate modification. MIT Press, Cambridge, Mass. 308 p.

de Sylva, D. P., and A. E. Hine. 1970. Ciguatera marine fish poisoning. A possible consequence of thermal pollution in tropical seas? FAO. Tech. Conf. on Mar. Poll. Rome. 8p.

The Institute of Ecology. 1972. Man in the living environment. University of Wisconsin Press, Madison. 267p.

Trott, L. B., and A. Y. C. Fune. 1973. Marine pollution in Hong Kong. Mar. Poll. Bull. 4: 13-15.

Van Bavel, C. H. M. 1966. Potential evaporation: The combination concept and its experimental verification. Water Resources Res. 2: 455-467.

Wade, B. A., L. Antonio, and R. Mahon. 1972. Increasing organic pollution in Kingston Harbour, Jamaica. Mar. Poll. Bull. 3: 106-111.

Webb, L. J. 1968. Environmental relationships of the structural types of Australian rain forest vegetation. Ecology 49: 296-311.

Witt, J. M., and J. W. Gillett. 1970. Pesticides, pollution and regulation, p. 199-203. *In* J. W. Gillett (ed.) The biological impact of pesticides in the environment. Oregon State University Press, Corvallis.

Wood, E. J. P., W. E. Odum, and J. C. Zieman. 1969. Influence of sea grasses on the productivity of coastal lagoons, p. 495-502. *In* Lagunas Costeras, Un Simposio. Univ. Nac. Autonoma, Mexico, D.F.

Zieman, J. C., and E. J. P. Wood. 1974. Effects of thermal pollution on tropical estuaries. *In* Pollution in the tropical marine environment. Elsevier, Amsterdam, in press.

Section 7

MECHANISMS TO SUPPORT AND ENCOURAGE RESEARCH AND EDUCATION IN TROPICAL ECOLOGY

Luis Montoya, Team Leader

Waldemar Albertin
Paulo de T. Alvim
Ana Luisa Anaya
Rufo Bazán
Douglas H. Boucher
Richard S. Davidson
Gladys de Tazán
Michael Dix
Manuel Elgueta
Ernesto Medina
Braulio Orejas Miranda
Roger Morales González
Roberto S. Ramalho
Miguel Revelo
Ramiro Rizzo Leal
Pablo Rosero
Jorge Soria
Les Whitmore

I. Introduction

The preceding reports describe the research questions in tropical ecology that are especially important from both a theoretical and a practical viewpoint. In this section, we will consider some of the institutional mechanisms necessary for the support of this research. Fully developed research programs in ecology are a relatively new feature in the scientific establishments of many developed countries. In most cases these programs have been stimulated by pollution problems, environmental deterioration, and needs for planning. Tropical American countries need ecological research since, as discussed in the other sections, it is probably not possible to directly transfer information from temperate regions to the tropics. Ecological research also underlies and directly supports efforts to expand food production and industry, control disease and pests, and effectively plan for the optimum use of the tropical lands and waters. However, some decision-makers have identified ecological research as antagonistic to these developmental projects. Some reasons for this attitude are a conservationism that aims to preserve natural ecosystems untouched by modern man, unreasonably sharp criticism based upon so-called ecological assumptions against carrying out economic developmental programs, and recently the emphasis on ecological research as a "solution" to environmental pollution and the maintenance of quality of life in industrialized countries, which simultaneously involve increased costs of industrial goods that may be purchased by less developed countries. Tropical American countries are facing numerous problems that have an ecological component. Whether these problems can be solved on existing developed lands with existing techniques or whether it is necessary to open more and more wild land is presently a much debated matter. The scientific basis for addressing these problems lies in a precise knowledge of the functional relationships of the biosphere. Thus, in a sense, mathematical models of population and community dynamics offer a real (but surely not the only) possibility for pest control; nutrient biogeochemistry in forest ecosystems offers a clue for understanding the loss of natural fertility when other communities are substituted for the forest and could allow the design of appropriate fertilization schemes or the development of adequate nutrient supply for managed plant cultures. In the same way, ecosystem analysis can help in the development of an integrated methodology of land use. There are many other examples of the application of ecological principles for a profitable and rational use of natural resources. It seems clear then, that tropical American countries might be able to develop projects of basic and applied ecological research of a predictive nature, which will allow a reasonable and objective evaluation of the conditions of their resources in the context of the social, cultural conditions appropriate to each nation and region. Tropical countries are in the position to avoid (through ecological research), the situation that developed countries are facing today in relation to their environment.

As shown earlier, ecosystems that are damaged or destroyed in the process of use or development recover at rates in tens to hundreds of years. Therefore it is of prime importance to begin the development of economic policies and resources in the context of local long-term stability rather than immediate goals. This is an enormous and difficult problem because the economic pressures are powerful and demand solution now and the kind of development needed will require cooperation among the Latin American countries at a level that does not now exist. An essential aspect of this cooperation should be the establishment of unified environmental laws and regulations applicable to national as well as international industrial development activities.

Finally, it is worth mentioning that there is an accelerated and frequently irreversible process of natural resource destruction due to irrational, commercial exploitation. This exploitation is unnecessary, because it does not follow plans of sound, long-term economic development and destroys potential resources which will be needed in the near future for more vital purposes. Some remarkable cases are the destruction of the piedmont and rain-green forests in the Andes and the forests of southern Mexico for wood exploitation, grazing or shifting agriculture. Other examples are the increasing destruction of mangrove forests in South and Central America and the disappearance of the high mossy and cloud forests in the Andes.

These challenges and problems can be approached through research, but effective research requires an adequate institutional base, trained workers, and integration into a communication network so that research done elsewhere in the tropics can be applied to local problems. Finally, research has to be linked to society's needs so that research results can be applied effectively. These problematic issues are considered in the following section.

II. Recommendations

1. Organization of research should be based on conceptual and practical priorities. The former indicates the more important gaps in ecological knowledge of the tropics, and the latter represents the urgent needs of tropical countries in such areas as agricultural development, land use, changes in food production techniques, industrialization, and pollution problems. Research groups should be organized so that basic and applied ecologists can coordinate their work. These groups should be developed at different tropical research sites and should have an international structure because of tho scarcity of adequate human resources in any single nation. To that end two not mutually exclusive alternatives are available: (a) formation of an independant coordinating office connected to some intergovernmental organization like FAO, UNESCO, or IICA. Such an office would have the task of encouraging research programs, obtaining funds for research and education in the tropics, facilitating and stimulating information transfer within the scientific ecological community, and establishing contacts with the public and the pertinent national governmental offices, and (b) creation of an interamerican institute of tropical ecology, which could serve as a training, coordinating, and information-collecting structure as well as the nucleus for an eventual specialized publication on pure and applied tropical ecology. Creation of such an institute should not require new physical facilities or infrastructure, but rather would use one of the existing educational or research centers of high level in the American tropics. Both mechanisms lead to the identification, utilization, and reinforcement of tropical research stations, which allow for the development of research projects in the most important tropical ecosystems.

2. Three levels to be considered in international cooperation in research are (a) agreements between countries, (b) agreements between institutions, and (c) agreements between researchers. Cooperation now functions on a very irregular and sometimes unacceptable basis. The following guidelines to improve present situations are

suggested: (a) Research projects funded or organized by nonresident scientists should incorporate local colleagues or advanced students in order to improve training and scientific quality of local groups. Funding agencies should insist that local cooperators be identified and assume that they will participate in research programs in a meaningful way, (b) Where local research groups exist, foreign scientists should be integrated with them and not operate independently, and (c) Foreign research groups should reinforce local institutions by giving advice to research programs and in every case, information obtained through their studies should be supplied to the host institutions either as reprints of papers published elsewhere, or in the form of articles published in a local journal.

3. The development of new funding mechanisms and sources will be necessary in order to allow Latin American scientists more independence from U.S. funded research programs. At present the U.S. National Science Foundation funding must be granted to and administered by U.S. citizens. Funds not restricted in this manner might be made available by foreign aid-granting agencies such as FAO, OEA, or UNESCO. Another possible funding mechanism would be the creation of an International Science Foundation, which could grant and administer funds for Latin American scientists.

4. Ecological studies of the American tropics are urgently needed in order to surmount problems of food production and protection of environmental values. Training of local personnel and international cooperation may be best accomplished in the following way. Those few people active in ecological research in the neotropics should have the opportunity of further training by means of intensive short courses in general ecology and tropical ecology. Complementing these people should be graduates in the fields of biology, agronomy, forestry, veterinary science, and other sciences who study ecology at the postgraduate level at those temperate zone universities that allow the student to do a thesis on a tropical problem in his home country.

5. A unified curriculum in ecology for students of tropical American countries is recommended. This curriculum should be offered in a few centers of university-level education in tropical America, with the objective of training professional ecologists. Further appropriate specialization will depend on those aspects considered to be high priority. This curriculum intends to develop professional ecologists in from 4 to 5 years. Postgraduate training should be offered for those interested in research. It is also essential to improve development of other disciplines related to ecology in order to increase the efficiency of professional ecologists.

6. In order to develop a useful and efficient communication mechanism between researchers in tropical ecology, we suggest that the international coordinating unit previously proposed include an information center utilizing modern data-processing equipment and

techniques. The names of both local and international investigators, their addresses, a short description of their research projects, and an abstract of completed research would be registered in this center. Publications relevant to tropical ecology would also be recorded and abstracted. An investigator could request this information, obtain it quickly, and effectively utilize it both in planning his research and in his interactions with other research groups on international and national levels. The abstracts of the completed research could be published in a special journal. Technical assistance should be provided by the coordinating unit to the researchers in order to stimulate publication and communication of research results. Also, the international coordinating unit should stimulate those journals already established. These journals should have strong editorial boards in order to maintain the highest scientific standards in the field, sponsoring institutions should guarantee the publication of publishable scientific work performed in tropical scientific centers, and a series of reprints and bibliographical compilations should be produced and widely distributed.

7. In the interim before formation of a coordinating center and data banks, it is desirable to have a quarterly newsletter on research in progress and recently completed (prior to publication) in the tropics. This is of very great importance in avoiding a duplication of effort and facilitating exchange of ideas and procedures prior to completion of projects. This is especially critical when so little time remains to do many kinds of tropical studies. This newsletter should not attempt to review current literature in tropical ecology except when an individual wishes to draw attention to a particularly significant paper.

8. The greatest possible effort should be made to reach the widest audience, including decision makers involved in basic and applied ecology, using all possible audiovisual aids. This process would be facilitated by short courses on science and ecology for the press and audiovisual reporters.

9. Frequent international meetings should be held in the tropical regions and courses dealing with ecological problems should be implemented.

III. Present Status of Ecological Research in the American Tropics

In tropical American countries there exist a considerable number of institutions conducting ecological research (see Appendix 1 for some references). The number of persons interested in the subject is quite large. TIE has listed about 500 persons who have an interest in ecology and are resident in tropical America. In some cases (e.g., Mexico, Brazil, and Venezuela), projects are developed mainly by local research groups of high scientific caliber; in others (e.g., Costa Rica, Panama, and Puerto Rico) foreign groups, mainly from the United States of America, perform most of the basic research. The main constraints for development of local groups are funds and qualified personnel.

Results of research conducted by Latin American groups are mostly published in a great number of journals and bulletins of limited circulation, whereas research results of foreign groups are published mainly in North American and European journals, frequently not available to the institution where the research was carried out.

It can be observed, in addition, that there are insufficient interrelations between ecological research groups in the American tropics. There is no Spanish language journal of ecology nor a Pan American tropical society of ecologists; however, the first Latin American Botanical Congress held in Mexico in 1972 points the way toward fruitful interaction of scientists. Frequently, there is a duplication of effort among countries of limited resources, whereas foreign research groups do not interact with local groups, remain isolated, and sometimes create situations that do not favor international understanding (Budowski, 1972; Janzen, 1972).

If international aid is to be most useful to Latin American countries, it is essential that those countries have direct control over the selection of researchers who come to work with them. Finally, it is worth mentioning that little attention appears to be given by funding agencies of foreign countries to the urgent ecological problems that are faced by tropical American countries. Again this may be mostly due to lack of suitable communication between nations to clarify what problems are considered to be of top priority.

A. Comments on the Establishment of Multinational Programs

It is highly probable that the trend of increasing interest in tropical ecological research on the part of foreign scientists will continue. That is, there will probably be more and more extranational effort in the future. This effort represents a pool of talent, a benefit, a resource to be welcomed and used, from the point of view of the tropical host country. Indeed, this effort is badly needed. There exist, for example, urgent problems such as pest control that call for intensified research in population dynamics and animal behavior. Changes in natural fertility when disturbances are exerted upon ecosystems may strongly impede recovery and succession and require research in ecosystem analysis. Also productivity problems, waste disposal and processing in tropical ecosystems, and many other topics require the attention of multinational research groups.

However, new thinking, both in the technique and attitudes of the visiting scientists, is urgently needed if the causes of science, the scientists, and the host region are to be optimally satisfied. For example, basic research is urgently needed in the tropics, but applied research is even more urgently needed. A balance between basic and applied research is required, which we admit will represent a drastic change in present trends and attitudes. The separation between applied and basic research, although more justified in developed regions, is not justified in countries faced with predevelopment problems.

It is vital for optimum development of ecological research in the tropics that Latin American talent be developed to a level much higher than it is now; in reality, most tropical American research should ideally be done by Latin Americans, in cooperation with foreign scientists when necessary.

B. Analysis of Past Efforts of Foreign Scientists Studying Tropical Areas

A review of research efforts of foreign scientists in tropical America reveals both strongly positive and negative results. Some general examples of the kinds of problems encountered are discussed below.

Some of the training programs in basic biology and ecology have professed the intent to train resident students and professionals, yet have actually focused entirely on extranational professionals from temperate regions. There is no question that it is of great benefit to tropical countries that residents and decision-makers in temperate regions understand the tropical environment and biota. However, courses on appreciation of the tropics should be clearly identified as such and not couched in terms of cooperative international programs. Further, it is decidedly more important to develop scientific techniques. This need has higher priority than the courses in appreciation of the tropics.

Training courses should be developed for the professional educator so that these resident specialists can in turn create the courses and workshops for their nationals. We can not expect that extranational professionals will fully appreciate or be able to work effectively within the subtle relationships of academia and government characteristic of each country.

Graduate students from temperate and tropical regions should be encouraged to spend part of their training period in other regions. These extended study visits will have to be supported by the government, and we urgently request increased funding in addition to those fellowship programs now in operation. Students from the tropics should be encouraged to do thesis research work at home so that the data gathered, the insight into local environments, the research tools developed by them can be put to use after they complete their degrees.

Temperate zone students also should work under the direction of appropriate specialists in the tropical region. In this way they would have a sponsor and avoid the problems associated with competition with local students and would gain in a social-cultural sense. Finally, they would be more able to operate within the local constraints in their research and not give the impression of unprincipled collectors of fauna and flora or competitors.

Research programs by extranational scientists have often approached tropical America in a spirit of exploration and discovery, ignoring the fact that although there are larger areas of underdeveloped land in the Americas, the societies of these countries are modern and developed. Scientists expect that they can follow the footsteps of the great explorers and begin research on any ecological topic in a virtually virgin field. This is no longer true. Almost every country in Latin America has trained a number of active young research workers in the biological and applied biology fields who look to their landscape as the source of research problems for their lifetime and that of their students. Today any visiting scientist must approach these local specialists and cooperate with them. The equipment, training, and vantage point of the visitor may be especially valuable and usually will be welcome.

Further, it is necessary to realize that science is carried out for a variety of reasons. In certain countries, research (often research publications) are the basis of the reward system. Both tropical and extra tropical cooperation must face these realities squarely and accept the fact that relatively rapid analysis of data and publication may be essential to maintain a relationship. But also these publications must include reports in the local languages in the local journals.

Clearly, there is a severe problem of integrating research workers into another national setting with different cultural attitudes toward research and different national priorities. These problems can be resolved by open discussion and interchange; in a spirit of cooperation and humility scientists should themselves develop the proper relationships and avoid social control because science depends upon the free exchange of information and ideas.

The final problem discussed under research concerns funding. As we have shown in several sections of this report, ecological problems transcend national boundaries. However, few nations have the resources or the interest to support research outside of their national boundaries. This means that research on regional problems or on biome processes is very difficult to support. Clearly, there is a need for interregional programs (such as IICA/TROPICOS) or international programs (such as UNESCO/MAB) which can stimulate and support regional programs. Further, it is essential that these be so organized that local concerns and loyalities cannot hamper the regional efforts. We see no obvious solution to this problem—it will require the continued attention of ecologists and ecologically minded administrators for a long time. Yet it is clear that this is the major reason we must maintain a vigorous exchange of researchers, students and other professionals between temperate and tropical countries.

IV. Education

A. Present Status of Educational Programs in Ecology in the American Tropics

There is presently in any tropical American country no formal curriculum for training professional ecologists. However, there are effective basic courses on general ecology designed for students of biology, agronomy, forestry, veterinary science, and others (see Appendix 1). No texts or other teaching materials presently exist that focus on the study of tropical ecological problems. The two principal ecology texts most widely used in Latin American countries are the following:

> Fundamentals of Ecology by E. P. Odum (in Spanish)
> Elements of Ecology by G. L. Clarke (in Spanish)

We suggest that there may be several schemes possible for training ecologists. The basic curriculum, however, requires a strong background in chemistry, mathematics, physics, and biology. Mathematics should include calculus and statistics, chemistry should include organic and biochemistry, and biology should include physiology, genetics, and taxonomy.

Besides this traditional science background ecologists need courses in population biology and genetics, energetics and behavior for those interested mainly in populations, and in geochemistry, hydrology, soils, vegetation, meteorology, geomorphology and systems analysis for those interested in communities and ecosystems.

An applied ecology program also must include work in the land-based sciences (agriculture and forestry), economics, and sociology. Clearly the integration of courses and programs between faculties and traditional schools poses difficult problems for the development of a sound program in basic or applied ecology. However, it is essential that this problem be solved by the educators and researchers if the impending environmental problems are to be resolved.

B. International Cooperation

Educational programs can, with international cooperation, be formed at three levels: (a) Privately initiated activity whereby visiting researchers, in groups or individually, offer seminars or consulting service to educational institutions in Latin America. (b) Agreements with universities in temperate zone countries that allow students to carry out thesis work in their (tropical) home countries in order to graduate from those universities. In this case, thesis work would be done in cooperation with local institutions such as experiment stations. Qualified extranational personnel should be encouraged to offer their talents and services where needed. (c) Participation of international experts in those places where the tropical ecology curriculum is implemented.

It is also appropriate that financial support for scolarships be provided to train students in ecology in foreign universities. Many Latin American countries do not have training programs in ecology and such financial support would provide training for needed personnel.

V. Documentation and Information in Tropical Ecology

Descriptive scientific information generated in basic and applied ecology in tropical American countries is published in several national and international journals specialized in those topics and in a variety of local journals (see Appendix 2 for some information). Scientific contributions in basic ecology are published mainly in international journals. The problem of retrieval and communication of these contributions is very large and it will be useful to researchers, students, and administrators of science to know which information sources are most likely to contain important papers or papers useful for a particular topic. Citation analysis (Garfield, 1972) is a technique to determine the most useful journals for a subject.

A citation analysis of the journals cited in the working papers used by the Tropical Ecology Workshop revealed that five journals were cited especially frequently. These were Journal of Ecology (40 citations), Ecology (36), American Naturalist (35), Science (31), and Turrialba (29). Of a total of 224 remaining citations, the frequency distribution of citations was cited once—136, twice—35, three times—19, four times—11, five times—6, six times—5, seven times—5, eight times—2, nine times—1, ten times—1, eleven times—2 and twelve times—1.

The papers cited by ecologists interested in the American tropics is very different for those journals considered most important in tropical and subtropical agriculture (Lawani, 1972). In that field Indian Journal of Agricultural Science, India Farming, Oleagineaux, and Journal of Economic Entomology were most important, where importance is based on the number of papers published. The only overlap between the most important journals and the most cited journals in this workshop were the Journal of Economic Entomology, East Africa Agricultural and Forestry Journal, and Turrialba.

Lawani (1972) also determined the language of the papers in tropical agriculture. About 73 percent of the papers in this subject are published in English, 8 percent in French, 8 percent in Spanish, and 7 percent in Portuguese. It appears from a cursory review of literature cited in this report that a similar pattern holds for American tropical ecology. This fact supports our conclusion that a Latin American journal of ecology is needed.

The analysis of ecology citations could be useful to librarians who have to determine which journals should be purchased with limited library funds. Clearly, the set of five most frequently cited journals should be in all laboratory libraries. We recommend that further citation research be done for tropical ecology since it is not only useful to librarians but also to students and scholars who must determine which journals to search for research data.

There has been a number of bibliographic studies of the literature for Latin American science, including ecology. These are listed in Appendix 2. The UNESCO, USDA, Library of Congress (USA) series of the official publications of American republics, and IICA bibliographies also provide general sources for library research. Besides these, most countries maintain bibliographies of papers and books in literature and science.

There is however, no Latin American journal for ecological research or ecological journal published in Spanish and Portuguese. Clearly there is a need for an international journal of high quality on tropical America or greater circulation of research results in the existent ecological journals. The language problems in the tropics cannot be avoided since they are obviously a severe barrier to communication. There will be a need in the future for tropical ecologists resident in the Americas to communicate directly with ecologists in Africa and Asia but this communication now is limited by language barriers and outlets for publication of research results.

APPENDIX 1

Sources of Information on Experiment Stations, Research Centers, Scientific Personnel, and Universities Concerned with Ecology and Environment in Latin America.

A. Research Institutions

The World of Learning. 1971-1972. 2 vols. Europa Publ. Ltd., London.

Master Directory for Latin America. 1965. M. H. Sable (ed.) Latin American Center, University of California, Los Angeles.

The Scientific Institution of Latin America. 1970. R. Hilton (ed.). Calif. Institute of International Studies. Stanford, California.

Directório de Científicos e Instituciones de Centro América y Panamá. 1968. Unesco, Centro Regional para el Fomento de la Ciencia en América Latina. Montevideo, Uruguay. 251 p.

Biological Research Centers in Tropical America. 1962. W. H. Hodge and D. D. Keck (eds.) Biological and Medical Sciences, National Science Foundation, Washington, D. C.

Biological Field Stations of the World. 1945. H. A. Jack (ed.) Chronica Botanica 9: 1-73.

Estaciones Experimentales Agrícolas de la Zona Andina: Bolivia, Colombia, Ecuador, Perú, Venezuela. 1970. Lima, Peru. 181 p.

Catálogo de Estaciones Agrícolas Experimentales del Istmo Centro-americano. 1971. IICA, Turrialba, Costa Rica. 104 p.

Estaciones Experimentales Agrícolas de America Latina. 1971. Inter-American Development Bank, Washington D. C. 367 p.

List of Marine Research Institutions in the Tropics. 1962. FAO, Rome.

Agricultural Research Index. 1969. Francis Hodgson Ltd., London. 206 p.

Directório de Instituciones Latinoamericanas de Oceanografia. Centro de Cooperación Científica para America Latina. Montevideo, Uruguay.

World Directory of Hydrobiological and Fisheries Institutions. 1963. AIBS, Washington.

Directório de Sevicios e Instituciones de Pesca en América Latina. 1963-1966. FAO, Santiago, Chile. 2 vol.

Fisheries Research and Educational Institutions in North and South America. 1950. FAO, North American Regional Office, Washington, D. C. 85 p.

Vademecum Forestal para América Latina. 1966. Oficina Forestal Regional de la FAO, Santiago, Chile. 105 p.

B. Scientific and Technical Personnel

Gutierrez, J. M. 1956. Lista provisional de técnicos agricolas de algunos paises de America Latina. IICA, Turrialba, Costa Rica. 78 p.

Instituto Interamericano de Ciencias Agrícolas, Oficina de Planeamiento. 1965. Latinoamericanos poseedores de grados avanzados en ciencias agrícolas. San José, Costa Rica. 72 p.

Molestina E., C. J., y M. T. Urizar M. 1968. Directório de profesionales en ciencias agrícolas de América Central. IICA, Zona Norte, Guatemala. Publ. Misc. No. 49. 155 p.

Fargas, J. E., J. Rea, C. E. Fernandez, y J. Leon. 1969. Investigadores agrícolas de la Zona Andina. IICA, Zona Andina, Lima, Perú. 283 p.

Mendes, A. J. T. Catálogos de especialistas en ciencias agricolas nos paises Latino-Americanos contendo un cátalogo dos centros de investigação agrícola da América Latina. Instituto Agronômico, Universidade de São Paulo, Cidade Universitaria "Armando de Salles Oliveira", São Paulo, Brasil.

Catalogo de Especialistas en Tecnología e Ciencias Agricolas no Brasil. Instituto Agronômico de Campinas.

List of Research Workers in the Agricultural Science in the Commonwealth and in the Republic of Ireland. 1969.

International Directory of Marine Scientists. 1970. FAO, Rome.

C. Colleges and Universities

Robles, L., J. Becerra,y F. Suarez de Castro. 1965. Estudio de la situación actual de las facultades de agronomía de Centroamérica. IICA, Zona Norte, Guatemala.

Seminário Internacional sobre la Enseñanza de Ecología y Suelos en las Facultades de Agronomía de América Central, San José, Costa Rica. 1967. IICA, Zona Norte, Guatemala. 120 p.

Caballero, H. 1968. Post-graduate institutions in Latin America: their objectives, problems and potentialities. In Panel on Post-Graduate Education and Associated Research for the Support of Livestock Development in Latin America. IICA, Turrialba, Costa Rica. 18 p. (Working paper No. A-1).

Franco Barbier, A.,y G. Naranjo. 1960. Directório de decanos y profesores de los centros de enseñanza agrícola superior en América Latina. IICA, Turrialba, Costa Rica. 124 p.

APPENDIX 2

Sources of Bibliographic Information on Science in Latin America.

Instituto Interamericano de Ciencias Agrícolas. Centro Interamericano de Documentacion e Informacion Agrícola.

a. Indice Latinoamericano de Tesis Agrícolas. 1972. Turrialba, Costa Rica. 718 p.

b. Bibliografia Agrícola Latinoamericana. 1966 to present.

Pan American Union. Division of Agricultural Cooperation. December, 1945.

Tentative Directory of Agricultural Periodicals, Societies, Experiment Stations and Schools in Latin America. 90 p.

USDA. 1943. A Preliminary List of Latin American Periodicals and Serials. Library List No. 5. 195 p.

Library of Congress U.S. 1958. A Guide to Official Publications of the American Republics.

Unesco. 1958. Directory of Current Latin American Periodicals. Pan American University, Paris.

Guide to Latin American Scientific and Technical Periodicals. Pan American Union, Organization of American States (OAS), General Secretariat, Washington, D. C.

Publicaciones Periódicas Bioagrícolas Latinoamericanas, Universidad de Narino, Instituto Tecnológico Agrícola, Pasto, Colombia.

References

Budowski, G. 1972. Scientific imperialism. 19 p. (mimeo).

Caceres, R. H. 1971. Publicaciones periódicas y seriadas agrícolas de América Latina. Una guia corriente de bibliotecologia y documentación, No. 19. IICA-CIDIA, Turrialba, Costa Rica. 69 p.

Garfield, E. 1972. Citation analysis as a tool in journal evaluation. Science 178: 471-479.

Janzen, D. H. 1972. The uncertain future of the tropics. Natur. Hist. 81(9): 80-94.

Lawani, S. M. 1972. Publicaciones periódicas de agricultura tropical y subtropical. Bol. Unesco Bibl. 26(2): 91-96.

INDEX